Contents

How To Use This Book

The ***Year 5 Targeting HASS Activity Book*** is filled with exciting stimulus materials to assist you in developing a greater understanding of the world you live in—its past and its present. The activities have been designed to encourage you to share opinions, think creatively, analyse ideas and information and respond in a variety of ways.

Each unit delivers content from the knowledge and understanding strand of the Australian Curriculum and supports the key inquiry questions in each subject area. The units also allow you to develop inquiry skills through:

Researching

Students identify and collect information, evidence and data from primary and secondary sources.

Questioning

Students develop questions about events, people, places, ideas, developments and issues to guide their investigations and satisfy their curiosity.

Analysing

Students explore information, evidence and data to identify and interpret features, patterns, trends and relationships, key points, facts and opinions and points of view.

Communicating

Students present ideas, findings, viewpoints, judgements and conclusions in digital and non-digital forms for different audiences and purposes.

Evaluating and Reflecting

Students propose explanations for events, developments and issues, draw evidence-based conclusions and use criteria to make informed decisions and judgements.

This book is organised into the Year 5 curriculum areas, as follows:

- **Units 1–12:** **HISTORY**
- **Units 13–24:** **GEOGRAPHY**
- **Units 25–28:** **CIVICS AND CITIZENSHIP**
- **Units 29–32:** **ECONOMICS AND BUSINESS**

TARGETING HASS 5 © PASCAL PRESS ISBN: 9781925726060

The *Targeting HASS Activity Book* series is designed to be a flexible learning resource. The units do not have to be completed in numerical order. Your teacher may direct you to complete units and activities from one learning area before moving on to a different learning focus. In this way, the *Targeting HASS* series will complement any school's scope and sequence.

The seven **Australian Curriculum's General Capabilities** listed below are embedded in the units of this book:

- Literacy
- Numeracy
- Information and Communication Technology (ICT) Capability
- Critical and Creative Thinking
- Personal and Social Capability
- Ethical Understanding
- Intercultural Understanding

A number of the activities and questions are open-ended. This allows for student challenge and differentiation. The questions are written to encourage you to think deeply about topics and provide extended responses. For some open-ended questions, answers are not provided. These questions are best graded through peers marking each other's work, or through student–teacher conferences. There are brief sample responses to guide the assessment of your work for some open-ended questions.

Analysing

Look at the images of the pioneering women. What do these images tell you about their lives? In what ways do the images support the text?

Evaluating and Reflecting

Are you and your family prepared for a natural disaster, such as a bushfire or flood? What would you need to do to better prepare yourselves?

The final section features eight **assessment** activities:

- 3 History
- 3 Geography
- 1 Civics & Citizenship
- 1 Economics & Business.

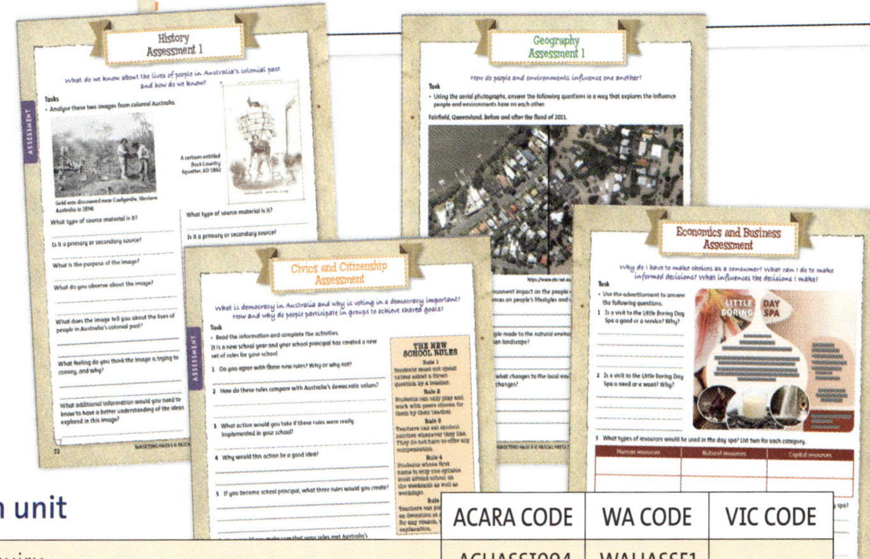

Each assessment activity relates to a key inquiry question. Make sure you have completed the units for that inquiry question before tackling the assessment.

Additional inquiry skills covered by each unit

		ACARA CODE	WA CODE	VIC CODE
Questioning	Develop questions to guide an inquiry	ACHASSI094	WAHASS51	
Researching	Locate and collect relevant information and data from primary and secondary sources	ACHASSI095	WAHASS52	VCGGC088
	Organise and represent data in a range of formats	ACHASSI096	WAHASS53 WAHASS58	VCGGC088
	Sequence information	ACHASSI097	WAHASS56	VCHHC082
Analysing	Examine primary and secondary sources to determine origin and purpose	ACHASSI098	WAHASS55	VCHHC083
	Examine different viewpoints	ACHASSI099	WAHASS57	VCHHC084
	Interpret data and information in a range of formats	ACHASSI100	WAHASS56	VCGGC090
Communicating	Present ideas, findings and conclusions	ACHASSI105	WAHASS61	
Evaluating and Reflecting	Evaluate evidence to draw conclusions	ACHASSI101	WAHASS59	
	Work in groups to generate responses to issues and challenges	ACHASSI102		
	Use criteria to make decisions and judgements	ACHASSI103	WAHASS60	
	Reflect on learning to propose action and predict the probable effects	ACHASSI104	WAHASS63	

Curriculum correlations

	History											
	1	2	3	4	5	6	7	8	9	10	11	12
ACARA CODE: DESCRIPTION – HISTORY												
ACHASSK106 Reasons (economic, political and social) for the establishment of British colonies in Australia after 1800			✔	✔								
ACHASSK107 The nature of convict or colonial presence, including the factors that influenced patterns of development, aspects of the daily life of the inhabitants (including Aboriginal Peoples and Torres Strait Islander Peoples) and how the environment changed	✔	✔	✔	✔	✔	✔	✔					
ACHASS108 The impact of a significant development or event on an Australian colony								✔	✔			
ACHASSK109 The reasons people migrated to Australia and the experiences and contributions of a particular migrant group within a colony										✔	✔	
ACHASSK110 The role that a significant individual or group played in shaping a colony												✔
ACARA CODE: DESCRIPTION – GEOGRAPHY												
ACHASSK111 The influence of people on the environmental characteristics of places in Europe and North America and the location of their major countries in relation to Australia												
ACHASSK112 The influence of people, including Aboriginal and Torres Strait Islander Peoples, on the environmental characteristics of Australian places												
ACHASSK113 The environmental and human influences on the location and characteristics of a place and the management of spaces within them												
ACHASSK114 The impact of bushfires or floods on environments and communities, and how people can respond												
ACARA CODE: DESCRIPTION – CIVICS AND CITIZENSHIP												
ACHASSK115 The key values that underpin Australia's democracy												
ACHASSK116 The key features of the electoral process in Australia												
ACHASSK117 Why regulations and laws are enforced and the personnel involved												
ACHASSK118 How people with shared beliefs and values work together to achieve a civic goal												
ACARA CODE: DESCRIPTION – ECONOMICS AND BUSINESS												
ACHASSK119 The difference between needs and wants and why choices need to be made about how limited resources are used												
ACHASSK120 Types of resources (natural, human, capital) and the ways societies use them to satisfy the needs and wants of present and future generations												
ACHASSK121 Influences on consumer choices and methods that can be used to help make informed personal consumer and financial choices												
NSW HISTORY STAGE 3 CURRICULUM OUTCOMES												
HT3-1 describes and explains the significance of people, groups, places and events to the development of Australia	✔	✔			✔	✔	✔	✔	✔	✔	✔	✔
HT3-2 describes and explains different experiences of people living in Australia over time	✔	✔						✔	✔	✔	✔	✔
HT3-3 identifies change and continuity and describes the causes and effects of change on Australian society			✔	✔	✔	✔	✔					
HT3-4 describes and explains the struggles for rights and freedoms in Australia, including Aboriginal and Torres Strait Islander peoples									✔			
HT3-5 applies a variety of skills of historical inquiry and communication	✔	✔	✔	✔	✔	✔	✔	✔	✔	✔	✔	✔
NSW GEOGRAPHY STAGE 3 CURRICULUM OUTCOMES												
GE3-1 describes the diverse features and characteristics of places and environments												
GE3-2 explains interactions and connections between people, places and environments												
GE3-3 compares and contrasts influences on the management of places and environments												

TARGETING HASS 5 © PASCAL PRESS ISBN: 9781925726060

Geography												Civics & Citizenship				Economics & Business				History Assessment			Geography Assessment			Civics & Citizenship Assessment	Economics & Business Assessment
13	14	15	16	17	18	19	20	21	22	23	24	25	26	27	28	29	30	31	32	1	2	3	1	2	3		
																						✔					
																				✔		✔					
																					✔						
✔	✔	✔	✔																				✔				
						✔	✔																				
				✔	✔			✔																✔			
									✔	✔	✔														✔		
												✔														✔	
													✔														
														✔													
															✔												
																✔											✔
																	✔	✔									✔
																			✔								✔
																				✔	✔						
																				✔	✔						
																						✔				✔	
																				✔	✔	✔					
✔	✔	✔	✔																				✔				
				✔	✔	✔	✔	✔	✔	✔	✔													✔	✔		
				✔	✔			✔	✔	✔	✔													✔	✔		

Curriculum correlations

	History											
	1	2	3	4	5	6	7	8	9	10	11	12
VICTORIA HISTORY LEVELS 5 & 6 CURRICULUM OUTCOMES												
VCHHC082 Sequence significant events and lifetimes of people in chronological order to create a narrative to explain the developments in Australia's colonial past and the causes and effects of Federation on its people								✔	✔			
VCHHC084 Describe perspectives and identify ideas, beliefs and values of people and groups in the past										✔		
VCHHC085 Identify and describe patterns of continuity and change in daily life for Aboriginal and Torres Strait Islander peoples, 'native born' and migrants in the Australian colonies											✔	
VCHHC086 Explain the causes of significant events that shaped the Australian colonies, contributed to Australian Federation and the effects of these on Aboriginal and Torres Strait Islander peoples and migrants								✔	✔	✔		✔
VCHHC088 The social, economic and political causes and reasons for the establishment of British colonies in Australia after 1800			✔	✔								
VCHHK089 The nature of convict or colonial presence, including the factors that influenced changing patterns of development, how the environment changed, and aspects of the daily life of the inhabitants, including Aboriginal and Torres Strait Islander peoples	✔	✔			✔	✔	✔					
VCHHK091 The causes and the reasons why people migrated to Australia from Europe and Asia, and the perspectives, experiences and contributions of a particular migrant group within a colony	✔	✔										
VCHHK092 The role that a significant individual or group played in shaping and changing a colony												✔
VCHHK093 The significance of key figures and events that led to Australia's Federation, including British and American influences on Australia's system of law and government			✔	✔								
VICTORIA GEOGRAPHY LEVELS 5 & 6 CURRICULUM OUTCOMES												
VCGGK089 Represent the location of places and other types of geographical data and information in different forms including diagrams, field sketches and large-scale and small-scale maps that conform to cartographic conventions of border, scale, legend, title, north point and source; using digital and spatial technologies as appropriate												
VCGGK091 Location of the major countries of Europe and North America, in relation to Australia and their major characteristics including the influence of people on the environmental characteristics of places in at least two countries from both continents												
VCGGK095 Impacts of bushfires or floods on environments and communities, and how people can respond												
VCGGK096 Environmental and human influences on the location and characteristics of places and the management of spaces within them												
VICTORIA CIVICS & CITIZENSHIP LEVELS 5 & 6 CURRICULUM OUTCOMES												
VCCCG008 Discuss the values, principles and institutions that underpin Australia's democratic forms of government and explain how this system is influenced by the Westminster system												
VCCCG009 Describe the roles and responsibilities of the three levels of government, including shared roles and responsibilities within Australia's federal system												
VCCCL013 Explain how and why laws are enforced and describe the roles and responsibilities of key personnel in law enforcement, and in the legal system												
VCCCC014 Identify who can be an Australian citizen and describe the rights, responsibilities and shared values of Australian citizenship and explore ways citizens can participate in society												
VCCCC016 Investigate how people with shared beliefs and values work together to achieve their goals and plan for action												
VICTORIA ECONOMICS & BUSINESS LEVELS 5 & 6 CURRICULUM OUTCOMES												
VCEBR001 Describe the difference between needs and wants and explain why choices need to be made												
VCEBR002 Explore the concept of opportunity cost and explain how it involves choices about the alternative use of limited resources and the need to consider trade-offs												
VCEBR003 Identify types of resources (natural, human, capital) and explore the ways societies use them in order to satisfy the needs and wants of present and future generations												
VCEBC004 Identify influences on consumer choices and explore strategies that can be used to help make informed personal consumer and financial choices												
VCEBC005 Consider the effect that the consumer and financial decisions of individuals may have on themselves, their family, the broader community and the natural, economic and business environment												

TARGETING HASS 5 © PASCAL PRESS ISBN: 9781925726060

Geography												Civics & Citizenship				Economics & Business				History Assessment			Geography Assessment			Civics & Citizenship Assessment	Economics & Business Assessment
13	14	15	16	17	18	19	20	21	22	23	24	25	26	27	28	29	30	31	32	1	2	3	1	2	3		
																					✔						
																					✔						
																				✔							
																				✔							
																						✔					
						✔	✔																				
✔	✔	✔	✔																				✔				
									✔	✔	✔														✔		
				✔	✔			✔																✔			
												✔	✔													✔	
													✔														
														✔													
												✔															
															✔											✔	
																✔											✔
																	✔										
																	✔	✔									✔
																			✔								✔
																			✔								

UNIT 1

Past Lives

The colony's first free settlers arrived on 16 February 1793 on the *Bellona*. They made their home at Liberty Plains, today's Bankstown. These settlers were told to bring an astonishing list of items because very little could be bought in the colony. Once they arrived, most settlers set up tents and began clearing land to build a slab hut like the one at right. The floor was usually mud, dirt or clay from a termite's nest, stamped down firmly. Sometimes boards were laid on the floor. Settler Emma Loader reminisced about settling in Coolgardie in 1896:

'The first thing … was to cut some posts and dig holes and put them in the ground, then cut corn sacks open and sew them together and nail them to the posts, then I put the iron on it for a roof and sewed more bags up and lined it inside too … I papered the inside with newspapers and whitewashed the outside, it was quite a smart house …'

Some of the colony's settlers spread out from the nineteen known counties of New South Wales. They established large grazing properties and became known as squatters. Squatting was not an easy existence in the early days of establishing a 'run'. They lived in tents while they constructed rickety huts made of roughly hewn timber and stringybark. They bound them together with rawhide, because nails were expensive. Sometimes wattle-and-daub huts were built by shaping moist mud over a framework of wattle twigs and then leaving it to dry. Hardwood shutters attached with leather hinges kept snakes and possums out of the homestead. Timber package cases and kerosene tins were converted to furniture and cupboards, and beds usually consisted of little more than sheepskins strung between springy saplings. Artist Alexander Denistoun Lang's painting from 1847 of a squatter's home is shown below.

In 1861, the *Crown Lands Act* allowed anyone to take up 'selections' from 4 acres to 320 acres for just 1 pound per acre. There was one condition: they must live on the land for three years with the intention of farming crops.

Staking a claim.

This opened up the bush to selectors, but caused ongoing struggles with the squatters who were reluctant to hand over parts of their properties.

Selectors were known as 'cockies' because they were often extremely poor and 'pecked up the grains of a living', just like a cockatoo. Their huts smelled bad because dish water, bath water and chamber pots were emptied close to the house. These smells combined with the smell of burning mutton-fat candles and cow manure, which was thrown on the fire at night to keep mosquitoes away.

In 1899, Steele Rudd wrote a book of short comic sketches called 'On Our Selection', which described the life of a selector. Rudd spoke about being so poor that they used burnt bread in hot water instead of tea. Meat was salted to make it last longer, or was kept in a 'Coolgardie safe', a wooden frame covered in wet hessian that cooled as the water evaporated. Rudd's book later became a stage show performed for modern audiences.

Source: Amazing Facts about Australia: *Early Settlers*, p. 19, Steve Parish

HISTORY

TARGETING HASS 5 © PASCAL PRESS ISBN: 9781925726060

Research

What did the first free settlers need to bring with them? What difficulties did they face in transporting these items to the colony?

Questioning

What parts of the text are primary historical sources?

How do you know?

What parts of the text are secondary sources?

How do you know?

Analysing

Is the 1847 painting of the squatter's hut a historically useful source? Explain the reasons for your answer.

Communicating

Using natural materials, create a miniature wattle-and-daub hut.

How successful was it?

Discuss your process and results with your class.

Evaluating and Reflecting

Which role would you have preferred in the colony: free settler, squatter or selector? Give reasons for your choice.

UNIT 2

Convict Life

HISTORY

Daily life

The lives of the convicts were strictly controlled. They were sent to Australia as punishment, so they were expected to work hard from sunrise to sunset. Male convicts cleared land for farming, chopped down trees for building, made bricks, and built roads and government buildings. Female convicts often worked as servants for officials or were sent to institutions known as 'female factories', where they completed daily tasks like laundry and cleaning the barracks.

Typical meals for convicts

Breakfast
Porridge made from oatmeal and water, bread

Lunch
Bread roll, dried and salted meat

Dinner
Bread roll

Clothes

This uniform from Van Diemen's Land was used in the 1830s and 1840s. It is known as a 'magpie' suit because of its half-black and half-yellow colour. It is made from woollen cloth and has an arrow design stamped on the trousers. The broad arrow mark is used to identify British Government property. This uniform was made to be worn by convicts who had been sentenced to chain gangs so they would stand out from other convicts. The outer parts of the trouser legs are fastened with buttons so the convicts didn't have to remove their leg irons when getting dressed. As well as the suit, convicts were also given two shirts, a sleeveless vest, one woollen cap and one leather cap.

Food

When the settlers first arrived in Australia, they had to bring supplies with them or have them shipped from England. By the early 1800s, a lot of the food the settlers ate was grown in Australia. In 1811, the weekly ration for convicts in New South Wales was 7 pounds of salt beef, 4 pounds of pork, 6 pounds of wheat and 15 pounds of corn.

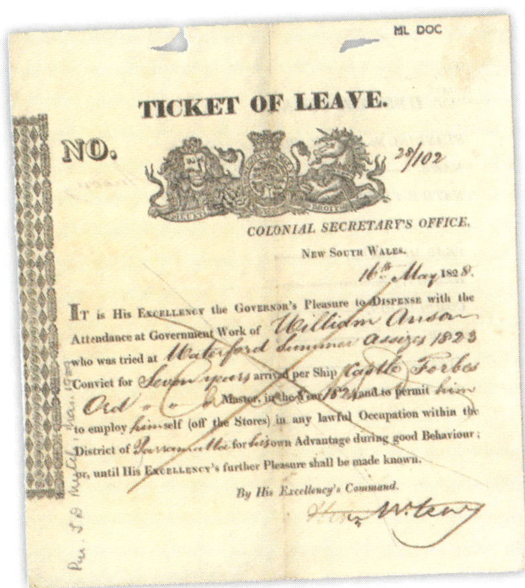

ML DOC

TICKET OF LEAVE.

NO. 28/102

COLONIAL SECRETARY'S OFFICE.

New South Wales.

16th May 1828.

It is His Excellency the Governor's Pleasure to Dispense with the Attendance at Government Work of William Anson who was tried at Waterford Summer Assizes 1823 Convict for Seven years arrived per Ship Castle Forbes [illegible] Master, in the Year 1824 and to permit him to employ himself (off the Stores) in any lawful Occupation within the District of [illegible] for his own Advantage during good Behaviour; or, until His Excellency's further Pleasure shall be made known.

By His Excellency's Command.

[illegible signature]

Punishments

If the convicts were badly behaved, they could be punished. Sometimes they were flogged with a whip called a cat-o'-nine-tails. They could be put into stocks or leg irons, or into a dark cell in solitary confinement. If the convicts continued to misbehave, they could be sent to a penal colony such as Norfolk Island or Port Arthur. Sometimes convicts were hanged. It paid to be a well-behaved convict, however. After four years of good behaviour, prisoners who had been sentenced to seven years' transportation were granted a Ticket of Leave, which allowed them to work for themselves. Some convicts were also pardoned and allowed to return to England.

Source: Australian History Centres: *Upper Primary*, p. 19, Blake Education

TARGETING HASS 5 © PASCAL PRESS ISBN: 9781925726060

Research

Find out about one of the penal colonies mentioned in the text: Norfolk Island or Port Arthur.

What was life like there?

Questioning

1 How many pounds are there in one kilogram?

2 What is a ticket of leave?

3 Why would you be given the cat-o'-nine-tails?

4 What did it mean to be given a pardon?

Analysing

What is the origin of the historical artefacts shown in the text?

Why have images of these artefacts been included?

What impact do the images of these artefacts have on the reader?

Communicating

Create a glossary of historical terms used in the text. Set it out in a table, with these column headings:

Term (word)	Definition in your own words	Picture or illustration

Evaluating and Reflecting

The crimes that convicts were being punished for included stealing a loaf of bread and being a vagrant (homeless). Do you think it was appropriate for convicts to endure such strict controls and punishments, given these crimes? Explain your answer.

UNIT 3

Growth of Australian Colonies: 1800s

HISTORY

By 1850, more than 187,000 free settlers had migrated to Australia.

Why did Britain want more people to migrate to Australia in the 1800s?

Food was in very short supply in the early days of the New South Wales colony. Very few convicts had farming experience and people did not know how to grow crops in the hot and dry Australian climate, as it was very different to the one back in England. The authorities wanted the British Government to send out farmers in exchange for land grants and free convict labour. This meant the British Government could colonise more areas of Australia and make money from agricultural exports.

Migrants embarking from Great Britain, 1880

Poster encouraging single women to migrate to Australia in the 1830s

Why did British people want to migrate to Australia at that time?

People wanted to come to Australia in order to find better lives for themselves. Life in Britain in the 1800s was difficult for many people. The Industrial Revolution put thousands out of work, as machines had been invented that meant fewer people were needed in factories. In Scotland, the Highland Clearances saw many farmers (crofters) evicted from their land so landowners could make large-scale sheep farms instead. Thousands of farmers lost their homes and needed to find new places to live. From 1845 to 1849, Ireland suffered from a Potato Famine where potato crops failed several years in a row. Many people relied on potatoes for their diet and livelihood. About a million people died as a result of the famine, and as many as two million may have migrated elsewhere.

How did the government encourage free settlers to migrate?

In the 1830s, the British Government began introducing assisted passage schemes for migrants, where the Government paid some or all of their fares. This gave poor labourers the opportunity to come to Australia. Through the Bounty System, employers in Australia used shipping agents in the United Kingdom to bring skilled labourers to the colony. If the migrants were deemed suitable by the Emigration Board, the employer or agent would be paid a sum of money (a 'bounty'). The assisted passage schemes were very successful in encouraging men to migrate, but the colony soon had far too many men and not enough women. Young married couples and single women were encouraged to move to Australia through the Bounty System. Between 1833 and 1837, about 2700 'bounty women' travelled to Australia.

Source: Australian History Centres: *Upper Primary*, p. 1, Blake Education

Research

Caroline Chisholm helped many 'bounty women' find work in the colony. Find out five interesting facts about her.

1 ______________________________

____

2 ______________________________

____

3 ______________________________

____

4 ______________________________

____

5 ______________________________

____

Questioning

Imagine you were a member of the Emigration Board in the 1830s.

Write three questions you would ask to find out if someone is a suitable candidate for the bounty program.

1 ______________________________

____

____

2 ______________________________

____

____

3 ______________________________

____

____

Analysing

What motivations did people have for migrating to Australia?

____

____

____

What type of person was most likely to migrate to Australia? Why?

____

____

____

____

Communicating

Imagine you were a farmer in the colony who needed skilled labour. Write a short advertisement to encourage people to migrate to the colony and work for you.

Evaluating and Reflecting

Was it the right thing to do to encourage single women, many of whom were orphans or poorly educated servants, to migrate to the colony when their future here was uncertain? Explain your answer.

____

____

____

____

____

Queensland 1859

HISTORY

The Governor of New South Wales, Thomas Brisbane, wanted to start a new penal settlement. It would be for convicts who had committed crimes *after* arriving in New South Wales. In 1823, he sent John Oxley, his surveyor, north along the colony's coast to find a site.

Oxley explored Moreton Bay and charted and named the Brisbane River. In 1824, Brisbane ordered the building of a penal settlement at Redcliffe. A year later it relocated to where North Quay is today. It was named Brisbane Town.

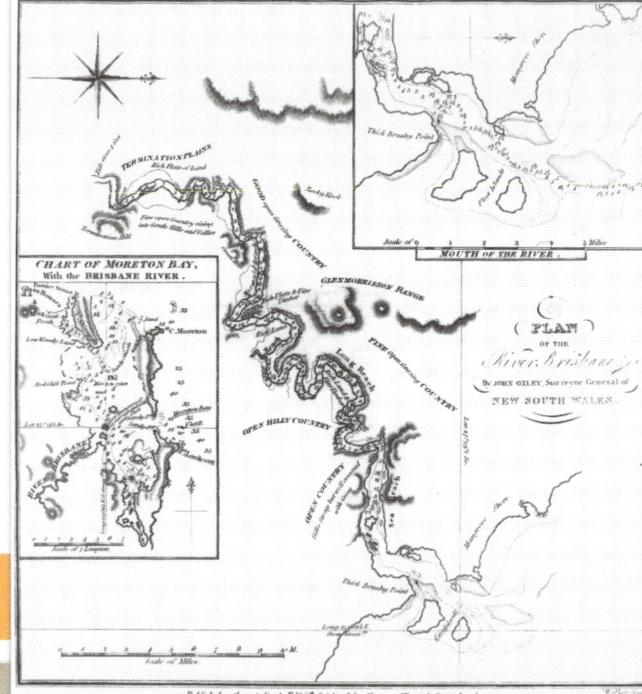

Oxley drew a detailed map of Moreton Bay and the Brisbane River.

Moreton Bay Settlement, 1835

Free settlers were banned from the area because the convicts were considered dangerous. The penal settlement was the harshest in the colony. The convicts were forced to construct buildings, roads and bridges. They cleared land for farms at Bundamba, Eagle Farm and New Farm.

When the Moreton Bay penal settlement closed in 1839, settlers moved in. The area grew very slowly, so free land was promised to British settlers. One group of Scottish people arrived on the ship *Fortitude* but found there was no land available. Some of them camped in the Bowen Hills. Their settlement grew into a large, busy community called Fortitude Valley.

The land around Moreton Bay was excellent for growing crops, particularly sugar cane which became a major industry. The area was renamed Queensland and became a new colony in 1859. By then, most of the land from the Fitzroy River (at Rockhampton) to the Darling Downs had been cleared for farming.

The Old Windmill in Brisbane was built by convicts in the 1820s.

Did you know?

An early name for Brisbane was Edenglassie, recognising the Scottish cities of Edinburgh and Glasgow.

Source: Go Facts Australia: *Colonies*, pp. 16–17, Blake Education

TARGETING HASS 5 © PASCAL PRESS ISBN: 9781925726060

Research

Find out three interesting facts about the Old Windmill in Brisbane.

1 ______________________________

____

2 ______________________________

____

3 ______________________________

____

What other significant sites in the area were built by convicts?

____

____

____

____

Questioning

Write four questions that you think would be answered by visiting sites built by convicts from the penal settlement, such as the Old Windmill.

1 ______________________________

____

____

2 ______________________________

____

____

3 ______________________________

____

____

4 ______________________________

____

____

Analysing

Look at the watercolour of the Moreton Bay Settlement painted by Henry Boucher in 1835.

In times when there were no cameras or mobile phones to record visual data, what types of information do sources like paintings and drawings tell us about life back then?

____

____

____

____

____

Communicating

Imagine you are John Oxley. Write a report to send to Governor Brisbane in New South Wales about your discovery of Moreton Bay.

Hint: Use the map provided in the text.

Evaluating and Reflecting

Should free settlers have been banned from Moreton Bay? Write a list of positive and negative outcomes if free settlers were allowed to live in the area.

Positive outcomes	Negative outcomes

UNIT 5

Colonisation and its Impact on Indigenous Peoples

Less than twenty years after James Cook first sailed into Botany Bay, the ships of the First Fleet arrived. They brought livestock, diseases and weapons, along with a mindset that the land was theirs for the taking.

LAND By taking land for farming, the European settlers deprived Aboriginal peoples of their previously abundant food and water sources, as well as their spiritual connection to their traditional country.

CONFLICT Without access to the native animals they once hunted for food, Aboriginal peoples often took the sheep and cattle that had replaced them. To settlers who believed in the idea of private property, this was an outrage and led to many conflicts.

Aboriginal peoples used their superior knowledge of the landscape to wage guerrilla and economic warfare on white settlers by killing livestock and burning property.

DISEASE Epidemics of European diseases – such as smallpox, chickenpox, measles and influenza – decimated Aboriginal communities.

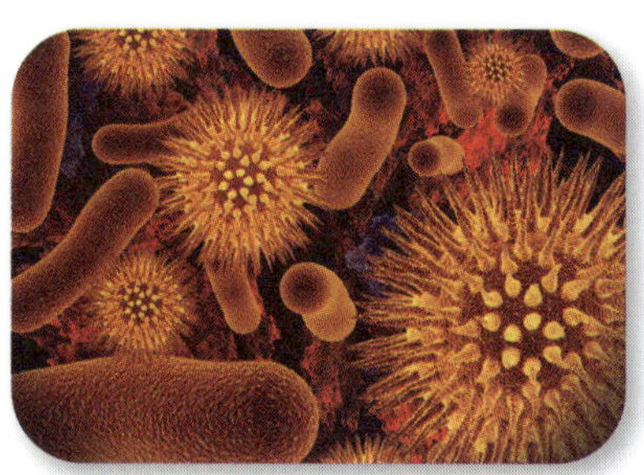

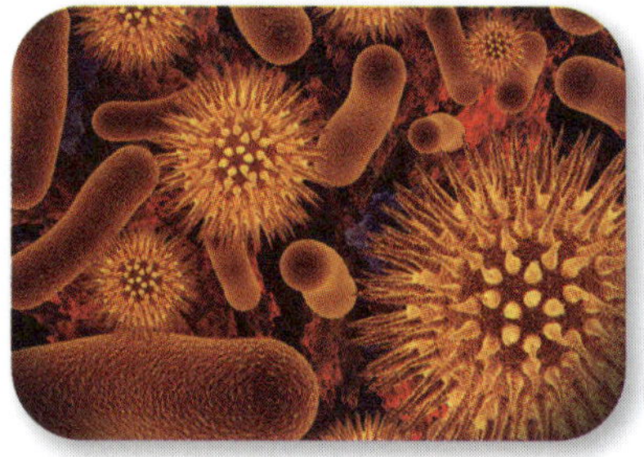

Aboriginal peoples had previously only consumed mild alcoholic drinks that were derived from traditional plants. It was the first settlers who introduced them to European alcohol and encouraged its consumption.

SPIRITUALITY Missions set up by religious organisations were intended to house and protect Indigenous peoples, and to convert them to Christianity. Little recognition was given to the traditional beliefs and customs of Indigenous peoples, and many did not feel 'at home' on missions.

POPULATION Between the years 1788 and 1900, it is estimated that conflicts and disease reduced the original Aboriginal population by as much as 90 per cent. By the early 1930s, the Indigenous population was estimated to be at its lowest – around 70 000. Over the next few decades, rates began to rise again as communities became better at adapting to life in a white-dominated Australia. By 1971, the estimated Indigenous population had doubled to around 150 000.

WAGES By the late 1800s, many Aboriginal peoples were living on land unsuitable for farming and were being exploited as labourers, often receiving food and clothing as payment. The government and other authorities would continue to control the wages of Aboriginal peoples until the 1970s.

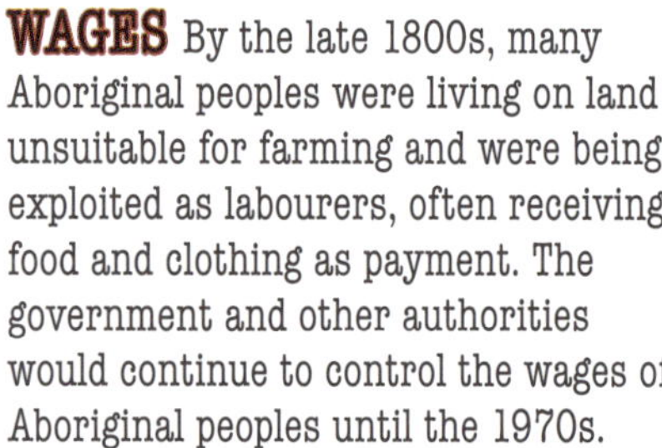

In 1984, a small group of Aboriginal people were found living in Western Australia's remote Gibson Desert. They are believed to be the last group to come into contact with white people for the first time.

Source: *Blake's Australian History Guide*, p. 38–39, Pascal Press

TARGETING HASS 5 © PASCAL PRESS ISBN: 9781925726060

Research

How did the colony of New South Wales grow from 1788 until 1820? What areas were explored and developed, and when?

Questioning

1 Why did so many Aboriginal people die from diseases such as measles and chicken pox?

2 Why do you think the government controlled the wages of Aboriginal people?

3 How did the attitudes of the Colonials influence their actions towards the Aboriginal peoples they encountered?

Analysing

In what ways did European colonisation change the traditional lives of Indigenous peoples of Australia?

Communicating

Using the information from your research, show the spreading impact of colonisation on a map of Australia.

Hint: Use a colour-coded key to show change over the years.

Evaluating and Reflecting

What advantages did the Aboriginal peoples have in their conflicts with early Colonials? Were the Aboriginal peoples within their rights to use these advantages? Give reasons for your answer.

Land Clearing and Introduced Species

HISTORY

One condition of owning land in the 1800s was that you had to clear it to grow crops, or to raise animals such as sheep and cattle. By 1860, 480 000 hectares of land were being used to farm crops. This widespread deforestation caused many native animals to lose their natural habitats, and several species became extinct, including the eastern hare-wallaby, the dusky flying fox and the broad-faced potoroo. There were laws against cutting down trees on riverbanks, but these were ignored. This led to soil erosion along many riverbanks.

Squatters occupied huge areas of pastureland for cropping and grazing. The number of sheep increased as wool became Australia's most important export. Overgrazing by sheep led to soil erosion and the loss of native plants.

The invention of the stump plough in the 1870s meant that areas previously unsuitable for farming could now be used. The plough made it possible to crop large areas of mallee (eucalyptus) scrubland much more quickly and easily.

Stump-jump plough in a paddock

The 1800s also saw increasing European migration to Australia. Some thought the Australian environment was strange and should look more like the one back in Britain. Acclimatisation societies were set up to import plants and animals to Australia to make the landscape look more familiar. Some of these imported species had a devastating effect on native plants and animals, and are still impacting our environment today.

One introduced plant was the prickly pear. It was brought to Australia to establish a cochineal dye industry. This red dye is created using insects found on the plant. But the plant spread very quickly, and by 1920, millions of hectares in Queensland and New South Wales were infested. It made the land useless for farming.

Today, if you visit the Northern Territory, you will see wild camels. They were brought to Australia from India, Africa and the Middle East during the 1800s to help explorers travel the interior of the country. The camels adapted well to the dry climate and were more resilient than horses. Camels were also used to help build the railroad between Port Augusta and Alice Springs. But by the 1930s, the use of cars and railways meant that camels were no longer needed. Many were released into the wild where they quickly multiplied. Camels eat many native plants, so they are seen as a threat to native species and their numbers are now controlled.

Source: Australian History Centres: *Upper Primary*, p. 19, p.21, Blake Education

Research

In 1859, rabbits were introduced into Australia to be hunted for sport. Find out what happened.

Questioning

What do the following words mean?

- acclimatisation:
- cochineal:
- resilient:
- export:
- infested:

Analysing

What impact would land clearing and soil erosion have on the Indigenous Australian peoples' connection with Country?

Communicating

Imagine you are representing an Indigenous group at a meeting of an Acclimatisation Society. They want to introduce a new species of fish into the Australian river system.

Prepare a 1–2 minute speech convincing them not to pursue this idea by outlining some potential hazards to people and the environment.

Evaluating and Reflecting

Today, camels are a pest. But when they were introduced, they performed a very important function in expanding and developing the inland areas of Australia. Overall, should camels have been introduced to Australia? What may have been a better way of managing the end of their role in transportation?

UNIT 7

HISTORY

Governor Macquarie

Lachlan Macquarie was born in Scotland and served as Governor of New South Wales from 1810 to 1822.

During his time as governor, the convict population of the colony increased rapidly as settlers demanded convict labour. Macquarie was a humane man who tried to fairly distribute prisoners while still maintaining public works.

He had grand visions for the colony, and as soon as he arrived, he set about making improvements. He banned pigs from being allowed to run in the streets, repaired damaged roads and falling-down buildings and stopped people from washing their clothes in the Tank Stream, which was the colony's water supply.

Macquarie initiated many of Sydney's fine constructions and his legacy remains in some of Sydney's most famous heritage buildings. A road was built to Parramatta, a lighthouse installed at South Head and factories and stables were constructed. In 1819, Macquarie built the first convict barracks and restarted government farms. The female factory at Parramatta was also rebuilt by Macquarie who thought it was "necessary for keeping those depraved females at work within walls, so as in some degree to be a check upon their immoralities and disorderly vicious habits".

NSW Parliament House, Macquarie Street, Sydney

One of the most pressing problems facing Macquarie as governor was the colony's food supply. He desperately needed to increase both agricultural production and the number of livestock. His solution came in expansion. He marked out new towns at Liverpool, Castlereagh, Pitt-town and Wilberforce. After Blaxland, Lawson and Wentworth found a route over the Blue Mountains in 1813, Governor Macquarie commissioned George Evans to plan a road over the mountains, to open up the prime grazing lands west of the Great Dividing Range. Once the road was completed in 1815, Governor Macquarie gave a full pardon to the 30 convicts who had carved out the 161 km route through the rough terrain.

Macquarie was known for his humane treatment of the convicts and many thought he was too lenient. Over his term, he freed 300 convicts on pardons and granted 1365 conditional pardons, while 20–25% of the prisoners arriving in Sydney were granted tickets of leave almost immediately. In 1819, the British sent JT Bigg to investigate Macquarie's management of the colony. Bigg was very critical of Macquarie's encouragement of financial enterprise among the convicts, because he thought it made transportation less of a deterrent. By 1821, Macquarie was ill and fed up with the criticisms. He resigned his position and returned to England the following year.

Source: Amazing Facts about Australia: *Early Explorers*, p. 49, *Early Settlers*, p. 50, Steve Parish

TARGETING HASS 5 © PASCAL PRESS ISBN: 9781925726060

Research

What public spaces in Sydney today were constructed during Macquarie's time as governor?

Questioning

What questions would need to be asked when planning the development of a public space, such as a park, garden, playground or library, in a large city like Sydney? Write four questions.

1

2

3

4

Analysing

What impacts would Governor Macquarie's colonial expansion have had on the local environment and Indigenous populations?

Communicating

Was Governor Macquarie a positive influence on the colony as a whole?

Discuss this question with a friend.

Do you have the same opinion and supporting reasons? Why?

Evaluating and Reflecting

What do you think would have been the most challenging aspect of Governor Macquarie's role: managing the convicts and free settlers, developing the infrastructure, improving food and water supplies, or expanding the colony by establishing new townships? Explain your answer.

Eureka Stockade

Colonial governments tried to establish law and order on the goldfields by banning alcohol, introducing gold mining licences, appointing a gold commissioner and sending police and troopers there. But riots and fights continued.

The Eureka Centre stands on the original site of the uprising on Eureka Street in East Ballarat.

NATIVE POLICE FORCE

Theft, alcoholism and fights were common and a police force to keep the violence and lawlessness in check was required. The first policemen on the goldfields were the Native Police Corps, a crack team of Aboriginal trackers who were excellent riders and hoped to track down criminals and bushrangers quickly. Miners called them "Joes", after their founder, Governor Joseph La Trobe. They were very good at their jobs but other Aborigines and Europeans began to resent them and they were disbanded in 1853.

A LICENCE TO REBEL

The colonies still needed farmers and the government hoped that by charging a licence fee, some of the less successful diggers would return to farming. Licences cost approximately 30 shillings (around $150) a month and diggers had to have their licence on them at all times, even when digging in mud or water that could ruin the paper licences. If miners couldn't immediately produce a licence, they were dragged off to prison or chained up by the police. Consequently, the diggers had no respect for the police and began to rebel.

A VOTE FOR THE EVERYDAY MAN

Trouble on the goldfields began in October 1854, when miner James Scobie was killed in a fight outside Bentley's Hotel, Ballarat. His friends believed hotel owner James Bentley was the killer and marched on the hotel, burning it down. Bentley was later found guilty of manslaughter and some miners were imprisoned for burning down the hotel. After the trial, citizens formed the Ballarat Reform League, which demanded the release of the convicted miners, the abolition of licence fees and a vote. Only landowners and squatters had the right to vote, but the diggers argued that they paid more for their licences than the squatters did for their leases, so the licences were effectively a land tax that should entitle them to vote. However, Governor Hotham refused to meet their demands, causing many diggers to publicly burn their licences and defiantly wave the Southern Cross flag.

On 29 November, after another licence hunt, fights broke out between troops and miners. The following day, Irishmen Peter Lalor (below right) addressed miners on Bakery Hill, urging them to build a stockade, swear an oath to "stand by each other" and fight the troops. By 2 December the stockade had been built. Unfortunately for the miners, the soldiers attacked at 4.45 am on Sunday 3 December, when many miners had left the stockade, believing they wouldn't be attacked on the holy day. The miners were outnumbered two to one and many were killed. Peter Lalor escaped with a wounded arm, which was later amputated, and the bloodied Southern Cross was torn down. Although the Eureka uprising was put down, it did bring change to the laws.

The FACTS!

VICTORIA HAD ONLY JUST become a colony, so Governor La Trobe had just 44 soldiers and 40 policemen available to him in 1851 – and all but two of his policemen were intent on quitting to seek their fortune!

MANY RACIST MINERS persecuted Chinese diggers, who were unfairly subject to special taxes and whose numbers were eventually restricted by the Victorian government. In 1857, 120 miners attacked the Chinese camp in the Ovens Valley and burnt their tents and temples. More than 2400 Chinese were killed, most drowning in the river as they hurried to escape the racist mob. From 1860–61, the violence escalated and bloody attacks killed more Chinese miners at Lambing Flat, the present site of Young, New South Wales.

EVENTUALLY, PETER LALOR was not only allowed to vote but was elected to Legislative Council, representing the goldfields district for more than thirty years!

Miners waved flags bearing the Southern Cross.

Source: Amazing Facts about Australia: *Early Settlers*, p. 53, Steve Parish

Research

What changes were made to laws on the goldfields after the Eureka rebellion?

What impact did these changes have?

Questioning

Imagine you could have a conversation with the wife of a miner who was at the Eureka Stockade. Write four questions you would ask her.

1

2

3

4

Analysing

In 1854, when the Eureka flag was first flown on Bakery Hill, the miners swore an oath.

How was the flag symbolic to the miners? What did it symbolise for the government?

Communicating

With a partner, discuss what life was like for colonial farmers and miners. Which role would you rather have and why?

Create a PMI (Plus, Minus, Interesting) chart for each role.

Evaluating and Reflecting

Should Peter Lalor have encouraged the miners to build the stockade and fight? Was there a different course of action he could have taken? Explain your answer.

HISTORY

Myall Creek Massacre

During colonial times, there were many conflicts between white settlers and Aboriginal peoples over land. Many atrocities against Aboriginal peoples were committed by gangs of white stockmen. One of the most significant of these events was the Myall Creek Massacre, which took place near Bingara in north-west New South Wales.

Illustration of the massacre from *The Chronicles of Crime* (1841)

On 10 June 1838, a group of twelve stockmen made up of convicts and former convicts rode to the cattle station at Myall Creek. Their intention was to kill any Aboriginal peoples they found. The men were led by a squatter named John Fleming.

The Aboriginal people at the Myall Creek Station were well known to the station workmen and owner. They were considered friendly and peaceful. When the stockmen led by John Fleming arrived at the station, they found a group of 28 Aboriginal women, children and elderly men sitting around their camp. The younger Aboriginal men were away working at a neighbouring station.

The stockmen rounded up the Aboriginal people and tied them together with ropes. They then took the group into the bush and murdered them. Two days later, the murderers returned to the site of the killings and burned the bodies.

The station manager, William Hobbs, was not at the station that day. When he returned and discovered the bodies, he reported it to the police. The men were arrested and put on trial. At the first trial in November, all the men were found not guilty. A second trial was held, and seven of the men were found guilty. They were hanged on 18 December 1838.

Although there were many terrible acts committed against innocent Aboriginal peoples, the Myall Creek Massacre was the first time that white men had been punished for the murders of Aboriginal peoples in Australia. It is also one of the most well documented of the massacres. Unfortunately, the violent conflicts between white settlers and Aboriginal peoples would continue on for many more years.

On 17 December 1838, a man from Sydney named J. H. Bannatyne wrote to an acquaintance in England about the incident. He described the details of the massacre and the sentencing that followed.

"By the bye, writing of sentiment, I must not forget to tell you of a circumstance which has agitated the public mind in the Colony lately ... A quantity of stockmen in the interior having had their masters' cattle speared by the Native Aborigines were determined on revenging themselves the first opportunity — falling in with a tribe of strange Blacks 30 in number — men, women & children, a tribe they apparently never saw before, totally innocent of the charge, for which they were slaughtered without having given the slightest provocation ... their trial created an extraordinary sensation in the Colony and will be the subject of gossip for many a long day yet."

Source: Australian History Centres: *Upper Primary*, p. 25, Blake Education

TARGETING HASS 5 © PASCAL PRESS ISBN: 9781925726060

Research

Another significant conflict between Aboriginal peoples and European settlers was the Battle at One Tree Hill in Queensland in 1843. What happened during the conflict? What similarities and differences were there between this event and the Myall Creek Massacre?

____

____

____

____

____

____

____

Questioning

Write three questions to help a partner analyse the illustration of the massacre.

1 ______________________________

____

____

2 ______________________________

____

____

3 ______________________________

____

____

Analysing

What impact does the extract from J. H. Bannatyne's letter have on the reader?

____

____

Is this extract a primary or secondary source?

____

Is it a reliable source?

____

What does it tell you about the mood of the community at the time?

____

____

Communicating

In a small group, discuss what may have motivated the stockmen to go to the cattle station at Myall Creek with the intention of killing any Aboriginal peoples they found.

Evaluating and Reflecting

Prior to the Myall Creek Massacre, there had been many terrible acts committed against innocent Aboriginal people. So why was this the first time Europeans had been punished for the murder of Aboriginal people?

____

____

____

____

UNIT 10

The Chinese in Australia

HISTORY

In the 1840s, there were no more convict transports from England. Farmers in New South Wales employed cheap contract labourers from China known as 'coolies'. The labourers worked for their Australian employers for several years to pay off the costs of their passages to Australia.

When gold was discovered in Australia in the 1850s, Chinese immigrants came, hoping to strike it rich. Over 40 000 Chinese people (mostly men) arrived in Australia between 1852 and 1889.

After the gold rush, Chinese settlers worked in market gardens or on farms. Others set up small grocery stores, became merchants, or worked as 'hawkers' selling fruits and vegetables in country towns.

Chinese gold diggers started gardening and growing fruits and vegetables on the goldfields. They set up 'cookshops' that attracted European diggers to their camps. When the Chinese moved to city centres, they started cookshops in areas with other Chinese businesses. These were the early versions of the Chinese restaurants that now exist all across Australia.

Chinese Arch, Melbourne, 1901

As immigration to Australia became more restricted, the Chinese population became more concentrated in cities. These areas of Chinese businesses, temples and restaurants eventually developed into Chinatowns. Melbourne's Chinatown was established in the 1850s and is the longest continuous Chinese settlement in the Western world.

Entrance to Melbourne Chinatown today

Chinese university students newly arrived in Fremantle, 1946

In 1901, the White Australia Policy made it more difficult for Chinese people to immigrate to Australia. Immigration policies became less restricted in the 1950s, but it was only after the White Australia Policy was abolished in the 1970s that significant numbers of Chinese people began to settle in Australia.

Source: *Blake's Australian History Guide*, p. 95, Pascal Press

TARGETING HASS 5 © PASCAL PRESS ISBN: 9781925726060

What were the significant events and who were the significant people that shaped Australian colonies?

UNIT 10

Research

Chinese immigration brought an increase in racial tensions on the goldfields. The worst event occurred in June 1861 at Lamming Flats in NSW. What happened there and why?

Questioning

1 Why were 'coolies' employed in the colony?

2 What impact did Chinese migration have on the environment?

3 What impact did Chinese migration have on the colony's economy?

4 Why did the Chinese population become concentrated in cities and larger towns?

Analysing

What challenges would early Chinese migrants have faced when arriving and settling in the colony?

Communicating

Imagine you are a Chinese migrant recently arrived on the Australian goldfields. Write a short postcard to your family, to tell them of your experiences.

Evaluating and Reflecting

What lasting impacts has Chinese migration had on the Australia we live in today? Have these impacts been positive or negative? Explain your answer.

UNIT 11

HISTORY

Afghan Cameleers

The 1800s in colonial Australia were a time of expansion and development. But when it came to crossing the arid central desert regions of the country, horses and donkeys were not suited to the harsh dry conditions. So instead, camels were imported, and with them came their handlers – the Afghan cameleers.

An Afghan camel driver with a camel train, 1911. (State Library of South Australia)

Despite their name, the 'Afghans' and their camels came from India, Pakistan and the Middle East. They were from different tribes and spoke many different dialects. Afghan cameleers were the first known members of the Islamic community in the South Australian outback. They played a significant role in establishing the religion of Islam in Australia, building the first mosque in Marree South Australia, another in Broken Hill and several in Western Australia.

Afghan cameleers worked in Australia on three-year contracts, carrying trade goods such as sugar, tea and wool. They helped to build communication lines for the overland telegraph between Adelaide and Darwin, as well as permanent transportation routes. This was integral in creating a permanent link between the coastal cities of the growing colony and the remote sheep and cattle grazing stations that produced much-needed supplies.

Across the centre of Australia, down the Birdsville and Oodnadatta tracks, the Afghans established rest-house outposts called *caravanserai*. The famous train that runs from Adelaide to Alice Springs and on to Darwin pays tribute to the legacy of the cameleers in its name – the Afghan Express, known commonly known as 'The Ghan'.

Through their widespread travels, Afghan cameleers developed ongoing relationships with local Aboriginal peoples. They exchanged skills, knowledge and goods with each other. Aboriginal people sometimes became guides for the cameleers, providing them with a better understanding of the Australian landscape.

There were a number of intercultural marriages between Aboriginals and Afghan cameleers, with many of their descendants still living across Australia today.

The Afghan cameleers had much in common with outback farmers and settlers, yet they still suffered from prejudice and discrimination. In 1895, a law was passed to prevent them from mining in the Western Australian goldfields after unproven allegations were made about Afghan people polluting water holes. Many were also refused entry back into Australia after visiting their homelands due to the 1901 *Immigration Restriction Act*.

But the usefulness of the camel ended with the development of the internal combustion engine. The camels were let go to run wild, and their Afghan handlers either returned home or stayed and established families in Australia. However, their cultural influence can still be observed today with the Camel Cup and date palm plantations of Alice Springs, Muslim mosques in Adelaide and mosaic artworks across central Australia.

Muslim mosque

TARGETING HASS 5 © PASCAL PRESS ISBN: 9781925726060

Research

Which colonial explorers used camels for transport?

Choose one of these explorers to investigate. How many camels did they use? Where did they explore? What was the outcome of their journey?

Questioning

Write four questions that check your understanding of the text. Use the starting words provided.

What

How

Why

Would

Analysing

Look at the picture with the camels.

Who do you think the white men standing in the centre might be? What makes you think this?

Communicating

You have been tasked with the job of leading a camel train from Alice Springs to Adelaide. Using a map of the area, mark the route you will follow. Remember to include features of the landscape, such as mountains, deserts and rivers.

Evaluating and Reflecting

Who benefited most from the work done by Afghan cameleers? What evidence do you have to support your opinion?

UNIT 12

HISTORY

Pioneering Women

A woman's work in the bush was never done. Most women worked from daylight to dusk, washing, tending the vegetable patch, keeping cows, chicken and sheep, baking bread, churning butter and setting cheese, as well as looking after their children and preparing and cooking meals for the family.

Women often had to keep the station or selection running while their husbands were away droving stock to market or attending to business in the cities. Pioneering women lived a lonely, isolated existence, and droughts, fires, floods, snakes and unassisted childbirth were just some of the trials and tribulations faced. Usually the nearest homestead was hundreds of kilometres away, so there was very little company for women. Respectable ladies would never be seen in a sly grog shop, so opportunities to socialise were few. Many women feared attack from the swagmen or itinerants who travelled the outback, seeking work. One pioneering woman, Emma Withnell, was left alone with her ten children on their station near Dampier, Western Australia for extended periods while her husband was working as a pearler. Her memoirs record how she one day became so afraid that someone was outside that she fired off a shot from a shotgun, only to later realise she had shot at her own hat and dress hanging on the line!

Through drought, flood and fire

Martha Cox, who farmed in western New South Wales in the early 1870s, recounted her experience of drought in her memoir. "Day by day the water went lower in the tank ... if rain did not come soon it meant ruin for there was no chance of shifting the poor sheep. Every green thing disappeared and only the dry grass seed, which the sheep licked up, stood between them and starvation ... Our great drawback was lack of vegetables because our tank water was too precious to use on a vegetable garden." Later, Martha and her husband moved to Wagga Wagga and Martha took over running the station when her husband died. A bushfire laid waste to 1400 sheep and burnt down her home, but Martha and her farmhands carried on, rebuilding the home and station. Other famous pioneer women were Jeannie Gun (of *We of the Never Never* fame), Miles Franklin, Eliza Fraser, Georgiana Molloy and Georgiana McRae.

A female first

One of the most important women in Australian history was journalist and novelist Catherine Spence (below right). She was just fourteen when she arrived in Adelaide in 1839. She became the first woman to write a novel about life in Australia, and Australia's first female political candidate. Like Caroline Chisholm, Catherine Spence was a social reformer who campaigned to help orphaned or poverty-stricken women and children and to improve education. She was a staunch campaigner for a more democratic electoral system, championing women's right to vote. Thanks in part to her efforts, and those of the Women's Suffrage League and the Women's Christian Temperance Union, in 1894 South Australia became the first place in Australia where women could vote. In 1897, when she stood (unsuccessfully) for the federal convention, Catherine Spence became the first female political candidate in Australian history.

Formidable feminists

Australia's determined female settlers did not just have to battle the bush, they also had to fight for their rights in a society that was largely male dominated. Until the late 1800s, women in Australia had no equality. Australian women were not entitled to vote until 1902, after Federation (although South Australian women received the vote earlier).

Until the 1870s, women were unable to own property unless they were single, because after marriage all of a woman's personal property belonged to her husband. Women were also unable to become lawyers or doctors, or study at university until the late 1880s. In the colony of New South Wales, women were also unable to take up land grants; although, if their father's salary was too meagre to support them, young single women were able to apply for a land grant as part of their marriage dowry. This was the government's way of providing an incentive for men and women to marry.

Eliza Walsh, who arrived in Sydney in 1819, helped change the rules of land ownership by enabling women to be granted land independently. She initially bought a small piece of land when she arrived in 1819, and applied to Governor Macquarie for more in 1820, stating that she had bought cattle worth £1000 and would buy another £1000 worth of stock were she granted more land. Her request was denied because the governor wrote that it was not fitting to "give Grants of lands to Ladies".

Eliza immediately wrote to the English secretary of state, complaining, "a Lady is able to conduct a Farm as well as a Gentleman". The secretary of state wrote back to Governor Macquarie, "I am not aware of any reason why females, who are unmarried, should be excluded from holding Lands in the Colony". Eventually, nine years later, Eliza Walsh was granted land at Paterson River in the Hunter Valley of New South Wales.

Source: Amazing Facts about Australia: *Early Settlers*, p. 24, Steve Parish

TARGETING HASS 5 © PASCAL PRESS ISBN: 9781925726060

Research

The text names Jeannie Gun, Miles Franklin, Eliza Fraser, Georgiana Molloy and Georgiana McRae as examples of famous pioneering women. Choose one to research. What role did she play in the colony?

Questioning

What questions do you still have after reading the text? What did it make you think about or wonder? Write three questions.

1

2

3

Analysing

Look at the images of the pioneering women. What do these images tell you about their lives? In what ways do the images support the text?

Communicating

What qualities did pioneer women possess?
What made them pioneers?
Deliver a short 1–2 minute oral presentation about the hardships faced by colonial women and how they overcame these challenges.

Evaluating and Reflecting

What do you think might have happened in the colony if pioneering women had kept to their traditional roles of cooking, cleaning and child rearing?

UNIT 13

Europe

Of the seven continents of the world, Europe is the sixth-largest continent in size. Only the Australian continent is smaller. Europe is located in the northern hemisphere. Asia borders Europe to the east, and Africa sits to the south. It is the third most populated continent behind Asia and Africa, and includes around 50 individual countries. Each country is different in size and they have different languages, cultures, climates, and human and environmental characteristics.

With so many countries in one continent, Europe is able to host many events where only countries from Europe, and others that are invited, are able to participate. These include the singing contest Eurovision and the European Football Championships. Some countries work together and share the same currency (Euro, €), and many are members of the group known as the European Union (EU). The EU is a bit like a club, and the governments of member countries work together to try and improve the lives of their citizens. Europe, in particular Greece, is the birthplace of western civilisation. It is also the continent where both World War I and World War II began.

GEOGRAPHY

EUROPE FACTS

Population: around 740 million

Languages spoken: more than 200

Smallest country (land area and population): Vatican City, a tiny country within the Italian city of Rome

Biggest country (by land area): Russia, also the biggest country in the world

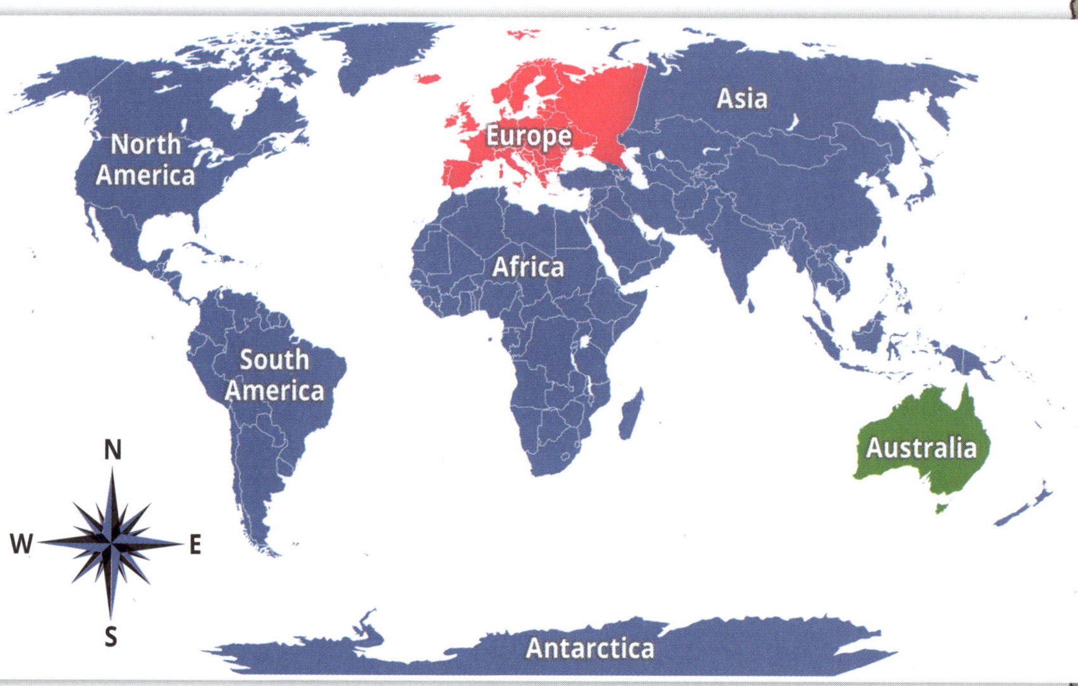

Human Characteristics

Constructed features such as buildings, roads and bridges, industrial processes and production (what is grown and made), communities and culture.

Eiffel Tower: France

Leaning Tower of Pisa: Italy

Parthenon: Greece

Channel Tunnel: Links United Kingdom to France

Stonehenge: England

Blue Mosque: Turkey

Environmental Characteristics

Naturally occurring features or processes such as climate, landforms, landscapes, soil, vegetation, water and minerals.

Highest mountain: Mt Elbrus, Russia. 5642 m above sea level.

Longest river: The Volga River, Russia. 3530 km long.

Wettest place: Crkvice, Montenegro. Average rainfall of 4600 to 5000 mm a year.

Driest place: Astrakhan, Russia. Average rainfall of 163 mm a year.

Coldest place: Ust-Shchuger, Russia. Once recorded a temperature of -55 °C.

Hottest place: Athens, Greece. Once recorded a temperature of 48 °C.

Source: Australian Geography Centres: *Upper Primary*, p. 1, Blake Education

TARGETING HASS 5 © PASCAL PRESS ISBN: 9781925726060

Research

Select one European country and write a brief information report about its people, climate and significant natural features.

Questioning

Use the text to help you write questions that have these answers.

Answer: hemisphere
Question:

Answer: Eurovision
Question:

Answer: Vatican City
Question:

Answer: Mt Elbrus
Question:

Analysing

Apart from their proximity to each other, what are some other reasons why countries in Europe cooperate and work together?

Communicating

Would you like to visit the European country you researched earlier?

Explain your reasons and give evidence by creating a travel blog post.

Evaluating and Reflecting

As time has passed, people have been able to travel further and more widely. In less than one day, a person can travel from Australia to Europe. In the 1850s, it took about four months! How do you think famous European landmarks have been affected by these developments in transportation?

UNIT 14

The Netherlands and Las Vegas

Humans have always changed their environment. Sometimes these changes are small. Sometimes they are so big they change the environment completely. Whenever humans make a change, they must ensure that these changes are sustainable. This means the changes they make will continue well into the future. This will only happen if the changes do not have a negative impact on the environment.

The Netherlands (means 'low-lying country')

The Netherlands is a small country in north-western Europe. The people who live there are called Dutch. The Netherlands is very low and flat. It is so low that around 26% of it sits below sea level and 17% used to lie under water. The Dutch people wanted more land to farm and live on, so they used windmills and large pumps to move the water away. The reclaimed areas of land are called polders. Water from the polders is pumped into waterways and constructed canals. These canals are also used for travel and irrigation. Large walls called dykes, and other barriers such as sand dunes, keep water from the North Sea out. Rising water from climate change is a threat to the Netherlands.

Windmills pump water out of polders and into canals and rivers.

Dykes stop the North Sea from flooding the land.

Las Vegas (means 'the meadows')

Las Vegas is a city in the United States of America. It began as a small settlement in the early 1900s. Today, it is one of the biggest cities in America and attracts millions of visitors each year. While the Netherlands has too much water, Las Vegas has too little. This is because it is located in the middle of a desert. A city in the desert can only grow if it has good access to water. Las Vegas gets its water from Lake Mead and from a series of underground aquifers. However, more water is being taken out than is being replaced by rain each year. It is expected that Las Vegas will become even drier and hotter with climate change.

Las Vegas glitter strip at night

The Hoover Dam stores water from the Colorado River in Lake Mead.

Source: Australian Geography Centres: *Upper Primary*, p. 5, Blake Education

GEOGRAPHY

Research

What was the Netherlands like before they began building dykes? Find out more about the process of 'land reclamation' and its impact on the local environment.

Questioning

What do the following words mean?

- aquifers:
- irrigation:
- reclaimed:
- canal:
- dam:

Analysing

Should the city of Las Vegas be allowed to grow and increase in population? Why? What are some positive and negative impacts of growth?

Communicating

Create a flow chart to explain how a windmill works.

Use diagrams with brief captions.

Evaluating and Reflecting

Should people try to change the natural environment by creating dams and dykes? Would it be better to leave the landscape in its natural form? Give details to explain your answer.

UNIT 15

The Trans-Alaska Pipeline

A Tlingit man of the Xunaa Kwáan tribe

The Inuit arrived in the Arctic region of Canada, Greenland and Alaska more than 4000 years ago from Siberia (Russia). Inuit means 'the people'. An Inuit person is an Inuk. (The term 'Eskimo' is no longer used.) Many Inuit go on camps during spring and summer, when they reconnect with the land, go hunting, and speak their traditional language. Family is very important to them, and Elders are treated with great respect for their knowledge and wisdom.

In 1968, oil companies discovered oil in the far north of Alaska. They wanted to build a pipeline to carry the oil south – across the land of Alaska Native tribes.

A protest against an oil pipeline on First Nations land. The Aboriginal peoples continue the fight for their rights.

The tribes did not want to lose their land but they knew the United States Government had the power to take it anyway to build the pipeline. They saw an opportunity to benefit from the project. Working together as the Alaska Federation of Natives, they tried to achieve the best outcome they could for their people.

Did you know?

The United States bought Alaska from Russia in 1867. The Tlingit and Haida peoples protested, arguing they already owned it.

The Federation agreed to exchange their land for US$962.5 million and 17.8 million hectares of land – about 12 per cent of the area of Alaska. They also agreed that the land allocated to the Trans-Alaska Pipeline could never again be claimed as Native American land. The Federation shared the money among the villages and regional corporations affected by the pipeline. These corporations then distributed shares to their local people. These shares continue to pay benefits to Native Americans, based on the profits made from Alaskan oil.

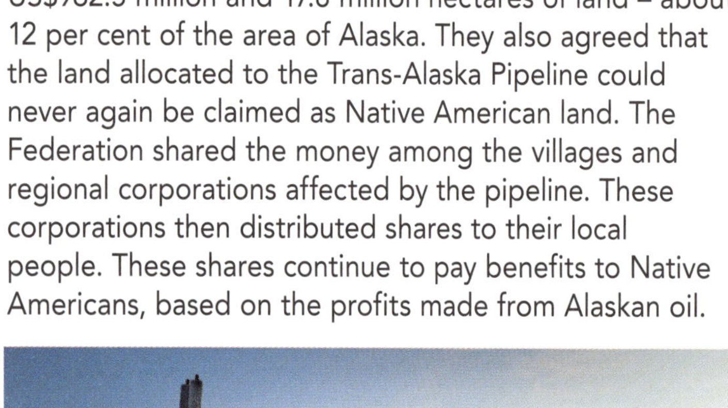

Good or bad?

Not every Alaska Native agreed with the pipeline deal. Some believe they both won and lost from its construction. Diane Benson, a Tlingit woman, said, "That thing cut through indigenous land a long ways. That can't help but have an impact directly on life, on hunting, on anything. Then on the other hand during those years we had a lot of poverty, you know, for Alaska Native people in particular."

Source: Go Facts Geography: *First Peoples*, pp. 10–11, pp. 14–15, Blake Education

GEOGRAPHY

Research

How do the Inuit interact with their environment?
What is their traditional lifestyle like?
Find at least five interesting facts.

____

____

____

____

____

____

____

____

____

____

Questioning

Write five questions about how the Inuit interact with their environment for a friend to research and answer.

1 __________

____

2 __________

____

3 __________

____

4 __________

____

5 __________

____

Analysing

Look at the images of the pipeline. List three ways people would have changed the environment during its construction.
Hint: What do you notice about the area around the actual pipeline?

1 __________

____

2 __________

____

3 __________

____

Communicating

With a partner or in a small group, set up a debate between representatives of the United States Government and The Alaska Federation of Natives to discuss the advantages and disadvantages of developing a second oil pipeline through the region.

Evaluating and Reflecting

What is more important: the traditional lifestyles of indigenous landowners and their connection with the land, or the need for energy sources such as oil? Is it acceptable to negatively impact the environment if it means providing energy for future development? Explain your answer.

____

____

____

____

____

American Dust Bowl

People can severely affect environments by the way they grow food.

During the 1930s, much of the United States experienced massive dust storms. They had two causes. First, for decades settlers across the country's Great Plains ploughed under the natural grasses. They planted the same crops they had planted in the wetter parts of the country, such as wheat. Second, the worst droughts in 1000 years struck and crops failed. The soil turned to dust, which the region's strong winds whipped into dust storms.

A dust storm approaching Stratford, Texas, on 18 April 1935. Some 770 million tonnes of topsoil blew away in this year alone.

GEOGRAPHY

Did you know?
Children wore masks and goggles to school during the dust storms.

From 1932, topsoil began to blow across several states. Each year the storms got worse. On 14 April 1935 – now known as Black Sunday – more than 20 storms combined to create one huge dust storm that turned the sky black. According to Herman Goertzen, a farmer from Nebraska, the chickens went to roost in the middle of the day because they thought it was night.

"It's important that we do the right thing by the soil and the climate. History is of value only if you learn from it."
– Wayne Lewis, Dust Bowl survivor from Oklahoma

Dust Bowl refugees headed west to find work.

Looking for work

The Dust Bowl's storms made the drought worse. They blocked the Sun, which reduced evaporation, cloud formation and therefore rainfall. The damaged land caused extreme poverty – no-one could grow crops or graze animals. By 1940, 2.5 million people had moved out of the Great Plains states to look for work. Most never returned. "The land just blew away; we had to go somewhere," said a preacher from Kansas. To make the situation worse, the dust storms occurred during the Great Depression, when unemployment was already very high.

Source: Go Facts Geography: *Reshaping Environments*, pp. 18-19, Blake Education

TARGETING HASS 5 © PASCAL PRESS ISBN: 9781925726060

Research

In January 2020, several major dust storms hit parts of Australia. Which locations were most affected?

What was the wider impact of these dust storms, in terms of the environment, human features of places and people's health?

Questioning

Write four questions that you would like to ask about the 'Great Plains states' of America.

1

2

3

4

Analysing

Look at the images from the text. What do they tell you about life in America in the 1930s, and the impact the dust storms had on the community?

Communicating

Create a labelled map to show the states that were impacted by the 1930s dust storms in America. Add captions using evidence from the text.

Evaluating and Reflecting

It is expected that dust storms will increase in frequency due to climate change. What are some actions that people living in these areas can take to reduce the impact of dust storms? Will these steps be enough? Why?

Responding to Climate

The climate of a place can affect where people live, their choice of house and what they do.

Coober Pedy is famous for its underground homes and churches.

The environment around Coober Pedy is dry and harsh.

Perth cityscape

Cottesloe Beach, Perth

GEOGRAPHY

Coober Pedy: Opal mining town

Coober Pedy is a small town in the north of South Australia. It is isolated and dry and has little access to transport, education or health services. People had little reason to live there until 1915, when opals were discovered in the area. Coober Pedy has a desert climate and is very hot during the day and very cold at night. But no matter what the temperature is on the surface, underground stays a comfortable 23–25 °C all the time, so people adapted to living in such harsh conditions by going underground. It is estimated that around 50% of the population of Coober Pedy live in underground homes called dugouts. There are also underground churches, shops and motels. Many people in Coober Pedy work in underground opal mines or in industries that support the mines.

Perth: Capital city of Western Australia

Perth is built on a large river (Swan River). When Europeans first arrived, rivers were used to transport people and goods, and for drinking and farming, so they were considered good settlement sites. The climate is another reason why many people live in Perth. Perth, like other coastal cities and towns, is much more pleasant to live in than hot, dry inland towns like Coober Pedy. People spend more time participating in outdoor leisure activities than they do in inland Australian towns. When the weather does get hot, people turn on air conditioners or cool off by going for a swim at one of the many beaches that Perth is known for. As the population of Perth increases, more jobs are needed to service it. This, along with the easy access to schools and health services, attracts more people to live there.

Average rainfall (mm) for Coober Pedy (1994–2020) and Perth (1993–2020) *Source:* Bureau of Meteorology

	JAN	FEB	MAR	APR	MAY	JUNE	JULY	AUG	SEPT	OCT	NOV	DEC	ANNUAL
Coober Pedy	13.7	14.1	10.7	12.3	9.8	12.4	4.7	6.5	8.9	10.2	13.8	19.1	133.8
Perth	19.1	13.4	19.2	36.0	86.8	127.9	144.5	125.5	82.8	38.8	21.7	10.6	733.2

Source: Australian Geography Centres: *Upper Primary*, p. 11, Blake Education

TARGETING HASS 5 © PASCAL PRESS ISBN: 9781925726060

Research

Coober Pedy is an opal mining town.
What are opals? How are they created?

Questioning

If you could talk to someone who lived in Coober Pedy, what three questions would you ask them about their environment and how it impacts their way of life?

1

2

3

Analysing

Using a Venn Diagram, compare and contrast the lives of people in Coober Pedy and Perth. What spaces would they have that were similar and different?

Coober Pedy

Perth

Communicating

Using the rainfall information in the table, create a graph to visually represent the variations between the months. You could choose to do one city or combine the two sets of data into a single graph.

Evaluating and Reflecting

Why do you think most Australians choose to live in places close to the coast, like Perth, instead of inland Australia like Coober Pedy? Give details to explain your answer.

Wooleen Station – Case Study

Farming in Australia has changed the environment and not always in a good way. Wooleen Station is a typical example of how some farming practices damage the environment. It is also an example of how this damage can be reversed.

CASE STUDY Wooleen Station is in Western Australia, 700 kilometres north-east of Perth. It was established in 1886, and the owners grew wealthy by running sheep and selling wool. In the 1930s, a long drought struck the area. This is when problems started to arise. With no rain, plants could not grow, yet large numbers of sheep continued to feed. In time, overgrazing damaged the environment on Wooleen Station. With little plant cover, erosion occurred as the fertile top soil was either blown away (wind erosion) or washed away (water erosion). In addition, the sheep trampled the bare soil so much that when the rain did come, the soil was so tightly packed together that rain could not soak into it, but instead pooled on top or ran off. Plants can't grow without soil or water, and the animals that depended upon them began to vanish. By the 1990s, Wooleen Station looked more like a desert than the lush green environment that first attracted farmers to the area.

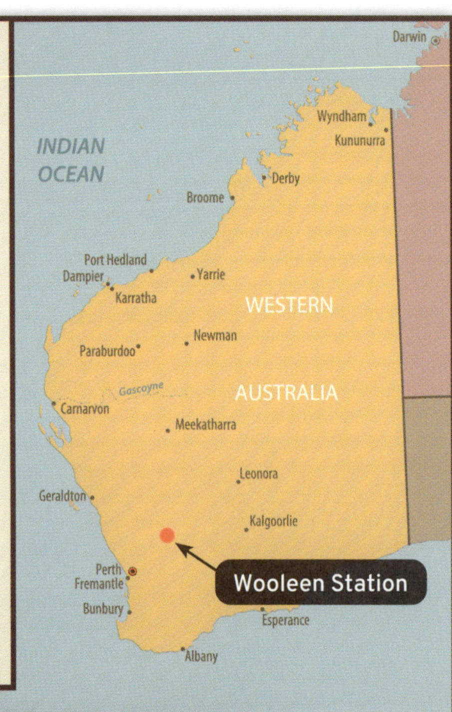

GEOGRAPHY

The owners of Wooleen Station today are working to regenerate the property back to its original glory. The water is cleaner; pastures and plants are growing back, and animals and birds are returning. They have done this by:

- destocking (taking all the livestock off the property to give it time to recover)
- fencing off important water sources to stop animals from trampling plants and muddying the water
- replanting trees and grass
- slowing down water flows from rainfall so the water seeps into the soil rather than runs off.

Overgrazing livestock has led to environmental damage on farms.

These photos show the recovery of the landscape on Wooleen Station over a period of three years.

Source: Australian Geography Centres: *Upper Primary*, p. 17, Blake Education

TARGETING HASS 5 © PASCAL PRESS ISBN: 9781925726060

Research

In 1985, Martin Royds took over Jillamatong, a family cattle farm near Braidwood in New South Wales where the old ways of farming had left the land in poor condition. What did Martin do to regenerate the land and begin farming sustainably?

Questioning

Imagine you were a government official tasked with analysing the impact of the changes made at Wooleen Station. What three questions would you ask the owners?

1

2

3

Analysing

Examine the before and after images of Wooleen Station. What impact has the recovery process had on the environment?

Communicating

Create a short speech or recorded podcast to convince other station owners in Australia to follow the example of the owners of Wooleen Station by regenerating their properties.

Evaluating and Reflecting

Once the regeneration of Wooleen Station is complete, should it be used to graze sheep for wool again? What should happen to it in the future? Give reasons to explain your answer.

Sustaining the Land

Aboriginal peoples' connection to country helps them to sustain the land.

For more than 50 000 years, the land gave Aboriginal peoples everything they needed – food, water, building materials, clothing and medicines. The way they worked with the land was guided by their spiritual connection to it. They had a duty to look after it. This responsibility is passed down in their law.

taro roots

Aboriginal peoples know a lot about the land, plants and animals. They use them carefully, so there is always enough food. For example, when collecting seeds they leave some behind for new plants to grow. When hunting, they don't kill young animals or mothers carrying young. They eat a variety of foods, so that no single food source runs out.

Torres Strait Islander people grow bananas, taros, coconuts and yams. After harvest, they leave the soil fallow to restore its health.

Europeans fear fire and Aborigines treat it as an ally.
Historian Bill Gammage

GEOGRAPHY

Did you know?
Looking after the land is called 'growing up the country'.

Patch burning of wetlands by the traditional owners (Bininj people) of Kakadu National Park, Northern Territory

Aboriginal peoples also lit fires to manage their country. Fire-stick farming is planned burning of the land to ensure food supplies. For example, people would burn the land in patches. Burning encourages fresh shoots and creates grasslands. This attracts animals, such as wallabies and wombats, which makes them easier to hunt. The unburned patches provide shelter for other animals. Traditionally, Aboriginal peoples burned early in the dry season. These fires are 'cool' and slow moving, which helps plants survive and gives animals a chance to escape. They also created firebreaks, which prevented uncontrollable fires later in the dry season. These traditional techniques of using controlled small, early season fires, emit much lower levels of greenhouse gases than big, late season bushfires.

Since the British arrived, bushfires have become bigger and more destructive. They threaten people, animals and property. They destroy biodiversity and damage the land. In the summer of 2019–20, there were fires across the eastern coast of Australia throughout Victoria, New South Wales and Queensland. Fires that started in separate states joined together and became 'mega-fires'. The World Wildlife Fund has estimated that 1.25 billion native animals died in these mega-fires.

Some plants cannot reproduce without fire. Heat and chemicals in the smoke cause seeds to germinate.

Source: Go Facts Geography: *People and the Land*, pp. 8-11, Blake Education

Research

Fish River Station is a large environmental reserve about 150 kilometres south of Darwin in the Northern Territory. How are they using traditional land practices to help manage the land and protect it?

Questioning

What do the following words mean?

- fallow:
- germinate:
- biodiversity:
- firebreak:
- sustainable:

Analysing

In what ways are traditional Indigenous land management practices sustainable? What might happen if Indigenous Australians don't manage the land sustainably?

Communicating

Create an artwork which represents the influence of traditional Indigenous land practices on the Australian environment. It might be a drawing or painting, a word or picture collage or a piece of digital art.

Evaluating and Reflecting

In 2019, Professor Hopper from the University of Western Australia suggested that all Australians should adopt a borrungur (Noongar word for totem) as a way of protecting native plants and animals and the environments in which they live.

Do you think this is a good idea? Would it work? Give reasons to explain your answer.

Indigenous Aquaculture

Not all Indigenous Australians were hunter-gatherers before the British arrived. They reshaped their environments to raise food.

GEOGRAPHY

Many early explorers such as John Batman and James Kirby found evidence of Indigenous aquaculture across the country that was designed specifically for the local area and fish species. For example, the Gunditjmara people of the Lake Condah region of Western Victoria farmed eels about 8000 years ago in the Budj Bim Cultural Landscape. This means that these eel traps are older than the Egyptian pyramids or Stonehenge. The Gunditjmara people used stones to build weirs, ponds and a series of channels to connect them over many square kilometres. Research suggests that their eel farm may have fed up to 10 000 people.

Further evidence of the detailed system was uncovered in January 2020 when a major bushfire burnt through parts of the Budj Bim National Park. As a result, archaeologists and local Indigenous rangers will do a cultural heritage survey of the site. This will help them find out more about the aquaculture system as well as the ancient stone dwellings nearby, which are believed to be Australia's only remaining permanent settlement built by an Indigenous community. In 2019, the eel trap site at Budj Bim was listed for its Indigenous heritage and cultural value with UNESCO (United Nations Educational, Scientific and Cultural Organisation).

Older still are the Brewarrina fish traps in north-west New South Wales, described by Bruce Pascoe in his book, *Young Dark Emu*. The traditional custodians of the fish traps are the Ngemba Wayilwan people.

Aboriginal fish traps in the Darling River, New South Wales. When water levels fall, the fish cannot escape.

Dreaming stories from the Ngemba people say the traps were the handiwork of the creation spirit, Baiame. Archaeologists calculate that the traps are at least 40 000 years old and possibly one of the oldest constructions on Earth. They were designed not only to catch fish, but also to allow some fish to pass through so that they could breed further up the river system, ensuring the trap's sustainability.

Over the years, disuse and deliberate destruction have led to the loss of many Indigenous aquaculture systems. Despite this, their significance in demonstrating the strong link between Aboriginal peoples and their Country remains.

Fish traps called Ngunnhu in Brewarrina, NSW, photographer unknown, c 1935.
https://sydney.edu.au/news/84.html?newsstoryid=1360

Source: Go Facts Geography: *Reshaping Environments*, pp. 6-7, Blake Education

TARGETING HASS 5 © PASCAL PRESS ISBN: 9781925726060

Research

What different types of fish traps did Aboriginal peoples use across Australia? What materials were they made from? How did they work?

Questioning

In your own words, what does 'aquaculture' mean?

Why were fish traps across Australia different?

How were the Brewarrina fish traps sustainable?

How does the existence of the fish traps show the Aboriginal peoples' spiritual connection to Country?

Analysing

Does this text challenge your previous understanding of Indigenous Australians and their traditional ways of life? How does it help you understand the way they managed their environment?

Communicating

Using your research, choose one type of fish trap. Create a labelled diagram to explain how it works.

Evaluating and Reflecting

The issues of commercial overfishing and the pollution caused by discarded fishing nets are huge problems that lead to decreasing stocks of fish and other marine life. Think about the sustainable methods used by Aboriginal and Torres Strait Islander peoples. What actions can individuals take to be more sustainable next time they go fishing?

UNIT 21

Town Planning

Humans organise spaces by creating zones. Similar activities are grouped together in the same zone. For example, people live in residential zones and they shop in commercial zones. The people who plan these zones are called urban planners or town planners. When deciding where to put each zone, planners take into account the landscape features of the area and the needs of the people who will use the zone.

Zone	What is found in the zone?	Notes
Residential	Homes: houses, apartments, townhouses, high-rise apartment blocks, etc.	There are often restrictions on the height, type and style of buildings in these zones.
Commercial	Businesses: shopping centres, petrol stations, offices, etc.	These zones often include a main street and are located near residential zones as people don't like to travel far to shop and work.
Industrial	Manufacturing and warehouses: transport depots, factories, warehouses, storage facilities, airports, etc.	These zones can be dangerous, noisy and smelly, so they are usually located a long way away from residential zones.
Green	Open land or parks: sports fields, gardens, nature reserves, etc.	These zones are often found near residential areas. Green spaces in the city are often small as land is expensive.
Special Purpose	Community services: hospitals, post offices, police stations, fire stations, churches, schools, etc.	These zones are often near residential areas as they provide services for the community.

Town planners follow laws made by state and local governments. These laws have to balance the needs of people with those of the government. This can be difficult, and often native animals like the endangered Southern Brown Bandicoot are the biggest losers. The newspaper article below is based on a true planning issue.

Town Planning Putting Wildlife at Risk

Wildlife corridors crucial to the survival of the Southern Brown Bandicoot may not go ahead if new planned suburbs proceed.

Plans for a residential development near the Cranbourne Royal Botanic Gardens have been revealed. These gardens hold one of the last remaining communities of bandicoots. Originally, the government had planned to include wide, fenced wildlife corridors to run from the gardens, through the suburbs to other areas of bandicoot habitat. However, these plans are now in doubt as the government thinks they are not cost effective.

The Southern Brown Bandicoot is listed as nationally endangered. If its last remaining remnants of habitat aren't protected, it faces extinction.

The wildlife corridors originally planned to run through the residential development would allow the bandicoots to breed in the gardens behind predator-proof fences, then travel safely via the corridors to other habitat areas. If the corridors are not included within the suburbs, then the bandicoots will be restricted to the gardens. This threatens the bandicoots' survival. If a catastrophe like a bushfire goes through the gardens, then all the remaining bandicoots will be lost.

The government will decide whether to include the corridors or not, and will release their revised development plans later next week.

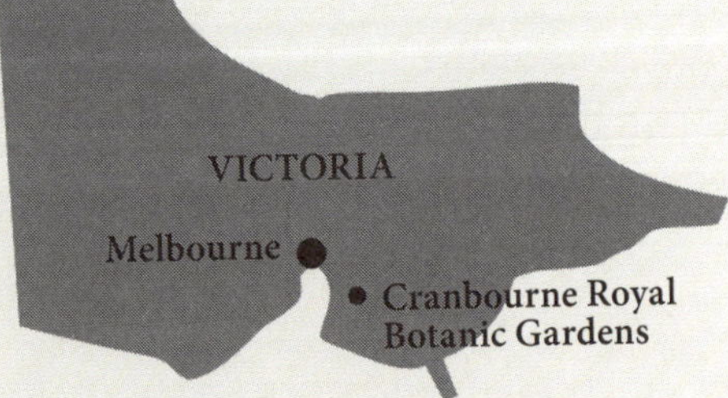

Source: Australian Geography Centres: *Upper Primary*, p. 21, p.23, Blake Education

TARGETING HASS 5 © PASCAL PRESS ISBN: 9781925726060

Research

What is a wildlife corridor? Find out where there are wildlife corridors in your state or territory. How are they managed?

Questioning

Use the text to help you write questions that have these answers.

Answer: industrial
Question:

Answer: zone
Question:

Answer: hospital
Question:

Answer: bandicoot
Question:

Analysing

What information does the table in the text provide? Is it useful? Explain your answer.

Communicating

Create a map of your suburb or local area. Use different colours to indicate different zones: Residential, Commercial, Industrial, Green and Special Purpose.

Evaluating and Reflecting

What are the advantages of organising spaces into zones?

Does your town or suburb have the right 'mix' of zones, or is there too much of one type and not enough of another? Give details to explain your answer.

Bushfires and Floods

Bushfires

Bushfires can occur all over Australia when temperatures are high and the environment is dry. During summer, different parts of Australia experience different weather conditions. In the north it is humid and wet, and in the south it is very hot and dry. Because of this, bushfires are generally more severe and more common in southern Australia, with most occurring during the hot summer months.

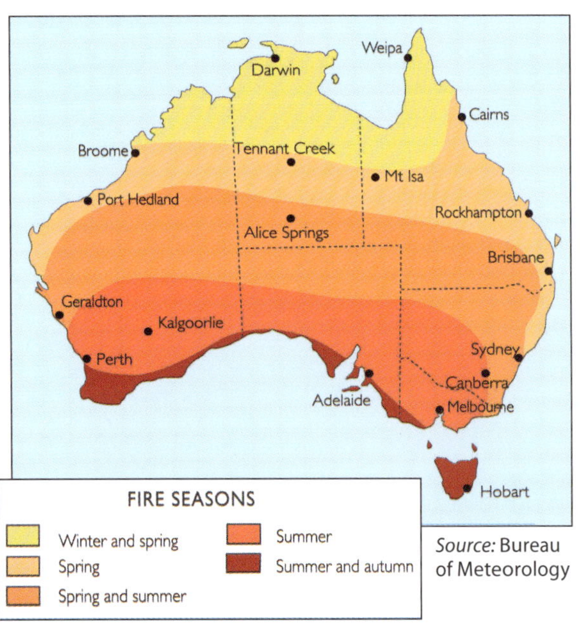

Source: Bureau of Meteorology

SIGNIFICANT AUSTRALIAN BUSHFIRES 2009–20

Date	Location	Deaths	Area (hectares)
2019 Nov–2020 Feb	New South Wales, Queensland, Victoria	34	18 626 000
2017 Feb	New South Wales	0	52 000
2016 Jan–Feb	Tasmania	0	100 000
2016 Jan	Southern Western Australia	2	70 000
2015 Nov	South Australia	2	85 000
2014 Jan	Victoria	1	55 000
2013 Jan	Tasmania	1	20 000
2009 Feb	Victoria (Black Saturday)	173	450 000

La Niña and El Niño are weather events that periodically affect Australia's eastern states. During a La Niña event, it is very wet and flooding may occur. During an El Niño event, it is very dry and severe bushfires may occur.

Flooding

Flooding can occur anywhere in Australia where there are creeks, rivers or water channels. Summer in northern Australia is called the wet season. This is when tropical cyclones and big tropical storms drop large volumes of water. As a result, most floods in northern Australia occur during summer. Southern Australia gets most of its rain during winter and early spring. For this reason, floods mostly occur here during winter.

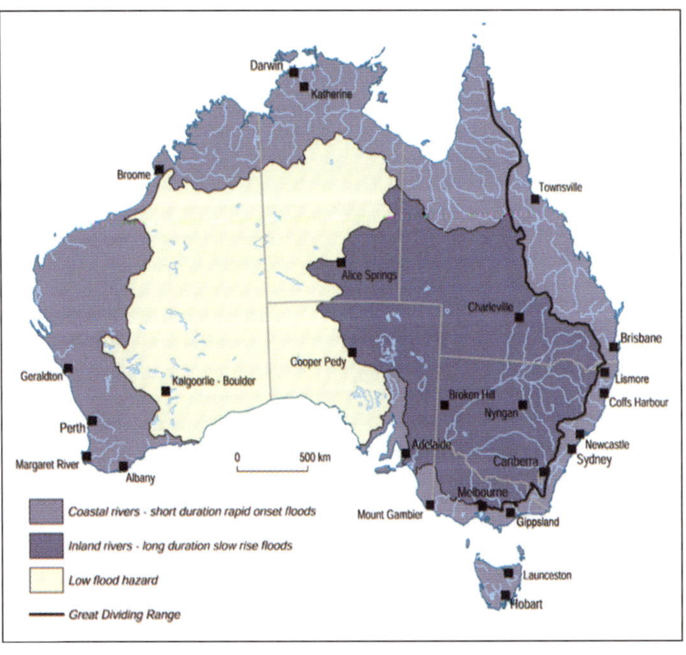

SIGNIFICANT AUSTRALIAN FLOODS 2010–20

Date	Location	Deaths
2020 Feb	Western Australia	0
2020 Feb	Sydney, Blue Mountains, Central West and Northern New South Wales	0
2017 Mar	Southern Queensland, Northern New South Wales	14
2017 Feb	Western Australia	2
2016 Sept	Central West and Riverina NSW	1
2016 June	Tasmania	3
2015 May	South-East Queensland	5
2013 Jan–Feb	Central and South-East Queensland, Northern New South Wales	6
2012 Feb–Mar	New South Wales, Victoria and Queensland	2
2011 Aug	Victoria	1
2010 Nov–2011 Jan	Central and South-East Queensland, Northern New South Wales	38

Source: Australian Geography Centres: *Upper Primary*, p. 25, Blake Education

TARGETING HASS 5 © PASCAL PRESS ISBN: 9781925726060

Research

Weeks after the devastating Queensland floods of 2019, water started flowing into Lake Eyre in South Australia. How did it get there? What makes Lake Eyre so special?

Questioning

Imagine you were developing a community program to limit the damage caused by fires or floods in your local area.

What questions would you need to answer before coming up with a workable plan? Write four questions.

1

2

3

4

Analysing

What information do the maps in the text provide?

What does the bushfire map show you about the link between bushfire timing and intensity and climactic conditions, based on location?

Communicating

Create a map showing the river systems that feed into Lake Eyre from Queensland. Label significant natural features: rivers, basins and mountain ranges.

Evaluating and Reflecting

Are you and your family prepared for a natural disaster, such as a bushfire or flood? What would you need to do to better prepare yourselves?

Reshaping Australia: Fire and Flood

Bushfires and floods are common in Australia.

Some bushfires and floods cause minor damage, but others cause deaths and widespread destruction. Two recent disasters had devastating effects on local communities.

On 21 November 2019, a lightning strike in East Gippsland, Victoria ignited a fire that began moving rapidly towards the coast. There had been little rain for months, so conditions were very dry and humidity was low. By 30 December 2019, there were three active fires in the Gippsland region, an area surrounded by old-growth forests.

Pushed along by strong winds, the fires reached the coastal town of Mallacoota at 8 am on 31 December 2019. Even though an evacuation order had been given for the region, almost 4000 people remained in Mallacoota, most of them tourists. Smoke from the fire billowed 16 km into the air as more than 16 tankers fought the blaze. Day turned to night as the smoke became so thick that it blocked out the sun. Fire crews were hampered in their work by falling power lines that sparked against trees and fences, and exploding gas bottles from household barbeques and local shops. The winds pushed the embers kilometres in front of the main body of the fire, which threatened to ignite dozens of spot fires.

On 3 January 2020, the Australian navy and *HMAS Choules* were called in to help evacuate 1160 people from Mallacoota's beach – the town was now cut off, with limited supplies of food, water and medical assistance. Victorian Premier Daniel Andrews declared a state of disaster. More than 780 000 hectares of land had been burnt out in the East Gippsland region. Mallacoota lost at least 100 homes and two people were confirmed to have been killed by fires. The blaze was finally contained on 20 February 2020, three months after it started.

The damaged environment in Victoria may not recover.

On 26 March 2017, category 4 tropical cyclone Debbie began strengthening off the Queensland coastline. Two days later, it made landfall near Airlie Beach and continued to travel south. Along the way, Debbie caused significant amounts of damage with sustained wind speeds of 175 km/h and gusts up to 215 km/h. Heavy rain, more than 1000 mm in 48 hours, inundated the Queensland and northern New South Wales coasts, causing heavy flooding. Fourteen people lost their lives and the overall damage was estimated to cost $3.5 billion.

Eastern Queensland experienced widespread power outages. There was major damage caused to the tourist hotspots of the Whitsunday Islands and Hamilton Island. Large sections of Queensland's sugarcane industry, up to 35% in some affected areas, were damaged. On Daydream Island, more than 300 tourists were left stranded because ships could not dock as the cyclone had destroyed the island's jetty.

Did you know?
Heat stress has killed more Australians than those lost in bushfires, floods, cyclones and storms combined.

Climate scientists predict that climate change will result in more and more severe natural disasters in Australia.

Prepare and Prevent

Disasters such as these lead to people wanting to know how to prevent or prepare for future disasters. The Victorian government announced that the Inspector General for Emergency Management (IGEM) would conduct an inquiry into the 2019–20 Victorian fire season. The inquiry will examine how well prepared Victoria was, and whether their response to the fire situation could have been improved. It will also examine relief and recovery efforts. The inquiry's first report is due by the end of July 2020, with the final report due the following year, June 2021.

Source: Go Facts Geography: *Reshaping Environments*, pp. 14-15, Blake Education

GEOGRAPHY

Research

Since the year 2000, how many major cyclones have impacted Australia?

Where did they make landfall, what path did they take and how strong were they?

Questioning

Imagine you could talk with bushfire or flood victims after they had recovered from their ordeal. What would you ask them about the impact the event had on their life and their community?

Write four questions.

1 ______

2 ______

3 ______

4 ______

Analysing

Look at the image of the damaged Alpine ash forest.
List four ways that bushfire has impacted this environment.

1 ______

2 ______

3 ______

4 ______

Communicating

Use your research to create a timeline to represent the frequency of cyclones in Australia.

Hint: Use a scale, such as 2 cm = 1 year.

Evaluating and Reflecting

It is expected that climate change will increase the frequency of severe weather events that cause bushfires and floods. How do you think this will impact Australia?

UNIT 24

Impacts of Bushfires

Bushfires affect human and natural features of the environment differently.

Natural Features: Vegetation

When an intense bushfire moves through an area of bush, it is left black and burnt. However, after a couple of weeks green growth appears. This is because many of Australia's native plants have adapted to fire. Some even depend upon it for survival. Plants survive by:

- using the fire to crack open fruit pods to release seeds
- re-sprouting after a bushfire.

Banksias, Hakeas and Eucalypts store their seeds in woody fruits. Smoke and heat from the fire opens the fruit. The seeds fall out and they germinate (sprout) in the nutrient-rich ash left after a fire.

Eucalyptus seed

Eucalyptus fruit

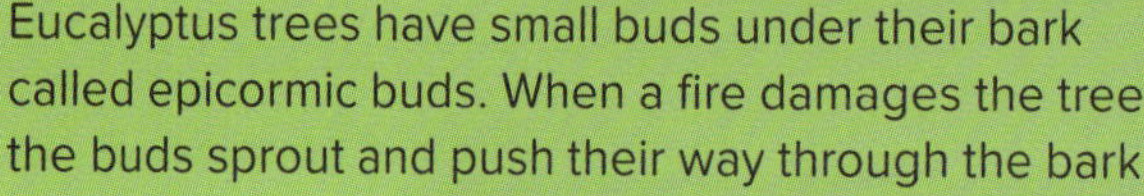
Eucalyptus trees have small buds under their bark called epicormic buds. When a fire damages the tree, the buds sprout and push their way through the bark.

Human Features: Communities

Hello, my name is Lucas. I live in Victoria and last year my town was destroyed by a bushfire. It burned everything, including my house, my school and my dad's butcher shop. It even destroyed roads, bridges and powerlines. My friend Sam lives on a farm and he relies on power to pump water from his tanks. When the power was cut, they had no water, so their house and sheds were burned too.

Straight after the fire, we were all in shock. Everything was burned. We had nothing left, but we had each other. We decided to stay, but many people left, including the baker and Doctor Wilson. We now have to travel one hour to the next town to go to the doctor or to buy fresh bread.

Our new house is almost ready and we should be moving in soon. In the meantime, we have been living in a caravan. Also, because my dad's butcher shop burned down, we didn't have a lot of money for a while. Dad has reopened and we're doing much better now.

Other businesses are slowly opening, and we're hoping a baker and a doctor will move to town soon.

When the fire came we left early, so we didn't see the worst of it. My friend Sam has bad dreams about the fire sometimes. Mum told me that's normal. She said it's a sign of trauma, and some people in town are affected like this. Sam is talking to a counsellor at school and he doesn't have as many bad dreams. Life is slowly returning to normal, but it is taking a long time.

Bushfires destroy homes.

Source: Australian Geography Centres: *Upper Primary*, p. 27, Blake Education

GEOGRAPHY

TARGETING HASS 5 © PASCAL PRESS ISBN: 9781925726060

Research

Each state has a bushfire service. These services not only help fight bushfires, they also provide the community with important information.

What is the name of your state's bushfire service? How can they be contacted? What steps do they recommend you should take in developing a bushfire plan?

____

____

____

____

____

____

____

Questioning

1 Which plants need fire to help them reproduce?

____

____

2 How is the ash from bushfires useful?

____

____

3 What are epicormic buds?

____

____

4 How can disaster victims recover from trauma?

____

____

____

Analysing

Show the impact of the bushfire on Lucas and his community, and what actions were taken after the fire.

Impact of bushfire	Actions taken after the fire

Communicating

Create a poster to show what actions people can take to minimise the impact of a bushfire on their homes.

Evaluating and Reflecting

After a major bushfire like the one Lucas experienced, many people in the town moved away. Should communities be rebuilt after bushfires, or is the danger too great with the risk of more bushfires in the future? What would you do – stay or leave? Why would you make that choice?

____

____

____

____

UNIT 25

Australian Democracy and Values

Democracy

Australia is a representative democracy. This means that the Australian public gets to vote for who they want to represent them in government. Australia's system of democracy was established when Australia became a federated nation in 1901. The rules that state how Australia's democratic system works are written in the Australian Constitution. Only a few of these rules have changed since the Constitution was first used in 1901. The changes that have most affected people's lives are those that concern who is viewed as an Australian citizen and how they are treated.

Values

The Australian democratic system is supported by several important values.

Freedom: Australians have the freedom to participate in elections and be elected. They have freedom of assembly (gathering to meet in public places) and the ability to be a member of a political party if they wish. Australians also enjoy freedom of speech and religious belief without discrimination.

Equality: All Australians have equal access to the legal system and civic rights. There should be no discrimination based on race, ethnicity, gender, age, marital status or place of residence.

Fairness: Linked to equality and justice, a 'fair go' enables all groups within Australia to be able to express themselves. It means supporting those who are disadvantaged, and it means being tolerant of the ideas of others and supporting inclusivity.

Justice: This is the rule of law. All Australians are required to uphold the law and all Australians are subject to the same legal processes. Laws must be fair and apply equally to members of the government, elected representatives and the general public.

Changes to the Australian Constitution and Laws

Federation – 1901

- Women are not allowed to vote.
- All Australians are classed as British citizens, not Australian citizens.
- Indigenous peoples are not allowed to vote in federal or state elections.
- Indigenous peoples are not counted in the census, and the federal government cannot make laws for Australia's Indigenous peoples — only states can.
- The White Australia Policy begins. Laws are created to only allow white, mostly British, people to move to Australia.

1902	1948	1962	1967	1970s
A law is passed to allow non-Indigenous women the right to vote in federal elections.	A law is passed stating that all people born in Australia are Australian citizens.	Indigenous peoples are allowed to enrol to vote in federal elections if they wish.	The Constitution is changed to include Indigenous peoples in the census and in any laws made by the federal government. They now have the same citizenship rights as non-Indigenous Australians.	The White Australia Policy is removed. People from many different countries migrate to Australia and become Australian citizens.

Source: Australian History Centres: *Upper Primary*, p. 51, Blake Education

CIVICS AND CITIZENSHIP

TARGETING HASS 5 © PASCAL PRESS ISBN: 9781925726060

Research

What changes occurred in 1901 when Australia ceased being a colony and became a federated nation? List the five changes that you think were most significant.

1 ______________________________

____

2 ______________________________

____

3 ______________________________

____

4 ______________________________

____

5 ______________________________

____

Questioning

Complete the chart.

What did I learn from the text?	What do I still want to know about Australian democracy and values?

Analysing

Australian democracy today is not the same as it was in 1901.

What trend can you see in the gradual change of the Australian Constitution and the laws around citizenship and voting?

____

____

What impact do you think this has had on Australian society?

____

____

____

Communicating

Should migrants have to become Australian citizens?

Discuss the question in a small group, and then list the pros and cons.

Evaluating and Reflecting

Which of our democratic values do you think is the most important, and why? How does this value impact on the lives of you and your family?

____

____

____

____

____

UNIT 26

Let's Vote!

Around 26 million people live in Australia today. Just under two-thirds of them are enrolled to vote. But how does the process work?

The Australian Electoral Commission

Also known as the AEC, the commission is an independent statutory authority. Its role is to conduct and supervise federal elections, by-elections and referendums, independently of the government. It also has the task of deciding how many of the 151 seats each state and territory will have in the House of Representatives at each federal election. This is done by using current population statistics. The AEC also keeps a record of all Australians who enrol to vote and the electorates where they live.

On the day of the election, AEC officials are there to assist people with the voting process and mark their names off on the electoral roll, but they cannot tell them who to vote for.

Secret Ballot

One of Australia's democratic values is freedom. This means people have the freedom to make their own decisions about which political representative they would like elected. To ensure this happens, Australia has a secret ballot. Once an AEC official has given the ballot papers to a voter, the voter is directed to a voting screen where they can number the boxes on the ballot paper in secret. Then they return the papers to the ballot box for collection and counting at the end of the day. The first secret ballots were held in 1856 in Victoria and South Australia.

Formal and Informal votes

When voting in an election, it is important to read the instructions and follow them carefully. Australia uses the preferential voting system for both the federal House of Representatives and the Senate. Voters number the candidates or political parties in the order that they prefer, starting with 1. If a ballot paper is completed in the correct way, it is a 'formal' vote and will be counted. However, if a ballot paper has been left blank, is incomplete, or has other writing on it, it is an 'informal' vote and it is not counted towards the election result.

Voting

Australia is one of only 10 countries in the world that enforces compulsory voting. If you are an Australian citizen aged 18 or older, you must be enrolled to vote. Once enrolled, you must vote in federal, state and local government elections. If you do not vote, you need to provide the AEC with a valid reason or pay a fine.

If you cannot get to a polling place on the day of the election, there are other options:

- Australians who are interstate, away from their own electorate, can vote at a specific interstate polling centre.
- Australians with a hearing or vision disability can use the AEC's telephone voting service.
- Australians who are overseas can vote before they depart (pre-poll), cast their ballot at an overseas voting centre, or apply for a postal vote.

How-to-vote cards

Political parties and their volunteers often supply voters with how-to-vote cards when they enter a polling place, particularly in marginal seats. This is a recommended voting order that is designed to support the political party's candidate. Unfortunately, some voters do not realise that they do not have to follow the how-to-vote card – they have the freedom to distribute their preferences in any way they like.

TARGETING HASS 5 © PASCAL PRESS ISBN: 9781925726060

Research

In Australian elections, there is a rule called the **election advertising blackout**. What is it? How does it work? What is exempt from this rule?

Questioning

What do the following words mean?

- statutory:
- referendum:
- electorate:
- ballot:
- marginal:

Analysing

One of Australia's democratic values is **freedom** – freedom of political expression and freedom of speech. Yet if a person chooses not to vote, they are penalised with a fine.

Is this fair? Should citizens be forced to vote?

Communicating

Create a mind map or word collage featuring key words and ideas to illustrate your knowledge and understanding of Australian democracy.

Evaluating and Reflecting

Should political parties hand out how-to-vote cards?
What are some of the positive and negative impacts of this strategy to gain votes?

Positive impacts of how-to-vote cards	Negative impacts of how-to-vote cards

UNIT 27

Enforcing the Law

Australia's democratic values include equity, fairness and justice. Our laws apply to everyone, regardless of their status, from the general public to judges and even the Prime Minister. The law can restrict the powers of governments, businesses and individuals, and people can only be punished if they are have broken the law. This 'rule of law' provides Australian society with stability and order.

Queensland park during the COVID-19 restrictions in 2020

Creating new laws in Australia is usually a lengthy process involving debate and a vote in both houses of parliament, followed by royal assent from the Governor-General. But at times, events change very quickly and the government must act quickly to protect people in the community.

The outbreak of the COVID-19 (coronavirus disease) pandemic in 2020 was a situation when the government had to act quickly and enact laws and regulations to stop the spread of the virus. They did this through a policy called 'social distancing'. Its purpose was to slow or stop the spread of this infectious disease. People in Australia were instructed to ensure that there was at least 1.5 metres of distance between them and other people. They also had to:

- stay at home unless absolutely necessary
- avoid greeting people with hugs, kisses and handshakes
- not visit other people
- practise good hygiene by washing their hands with soap and water for 20 seconds, and by sneezing into their elbows.

In addition to these health guidelines, the Australian government banned international travel. They closed down cafes, restaurants, gyms, libraries, community centres and swimming pools so that groups of people couldn't congregate in a small area. Many people lost their jobs when these businesses and others had to close.

They also limited the number of people who could attend a wedding (5) and a funeral (10), and even then, people had to stay at least 1.5 metres apart.

People who had already contracted COVID-19, or who were at high risk after being in contact with someone else who was infected, had to be socially isolated. They were required to quarantine themselves for a minimum of 14 days and they were not allowed to go out for any reason.

State police departments were responsible for enforcing the social distancing restrictions. In New South Wales, police could issue warnings to people who were not following the rules. They also had the power to issue on-the-spot fines: for an individual it was $1000 and for a business it was $5000. For worse or repeated breaches, police could arrest individuals and business operators who would then have to attend court and face a maximum penalty of $11 000 and/or six months in jail. After hundreds of people ignored the social distancing guidelines and flocked to Bondi Beach in Sydney, the government was forced to officially close the beach, blocking it off with red-and-white tape and using lifeguards and council rangers to enforce compliance.

In Victoria, the government declared a 'state of emergency'. This gave it the power to detain people, restrict the movement of people, prevent people from entering businesses and public spaces, and take other steps necessary to 'protect public health'. The Victorian government also had the power to quarantine entire suburbs, businesses or professions if it needed to. Many states, such as Queensland and South Australia, closed their borders and stopped people travelling from one state to another.

These protective laws and regulations were not meant to last forever. They would slowly be removed as the threat of infection from COVID-19 passed, after several months.

CIVICS AND CITIZENSHIP

TARGETING HASS 5 © PASCAL PRESS ISBN: 9781925726060

Research

Australian Border Force officers play an important role as a law enforcement agency.

What are they responsible for? What role did they play during the COVID-19 pandemic?

Questioning

Talk to a friend. What are their top three questions about the COVID-19 social distancing regulations?

Can you find the answers? You may need to do some additional research.

Question	Answer

Analysing

Why is it difficult to regulate people and their social interactions?

Communicating

Create a poster, radio infomercial or television announcement promoting community awareness of an important law or regulation.

Evaluating and Reflecting

When people break the law or disobey government regulations by ignoring social distancing rules, should they be punished? Why? Are there situations when punishment isn't necessary? What are they?

Getting Hands On

There are plenty of worthwhile societies and organisations in Australia that people can join to help others, to make a difference in their community, and to feel involved in a global cause. People don't get paid financially for the work they do, but they can gain valuable skills and experience, and give back to society.

Wildlife Rehabilitation

One of the best known wildlife care programs is run by **WIRES**, the **NSW Wildlife Information, Rescue and Education Service** (www.wires.org.au). They run training courses on caring for wildlife. Volunteer wildlife carers have to be registered and must understand the demands of fostering native Australian animals, some of which require round-the-clock care. Groups such as the **Royal Society for the Prevention of Cruelty to Animals (RSPCA)** and **Australia Fauna Care** (www.fauna.org.au) can also put volunteers in touch with qualified wildlife carers and organisations that can assist them with getting the training they need to be hands-on in caring for Australian animals.

The **Clean Up Australia** organisation (www.cleanup.org.au) has volunteers around the globe who work together to inspire individuals to clean up their local environments. They annually host Clean Up Australia Day and lobby for cleaner consumable products made of recyclable materials. More than 17 million environmentally conscious volunteers have picked up more than 365 000 tonnes of rubbish over the past 30 years as part of Clean Up Australia activities.

Joining a conservation organisation or becoming a volunteer firefighter might not be for you, but that doesn't mean you can't get involved in other ways. You can help by building nest boxes for animals in your backyard for native animals such as possums. You can encourage your school to limit plastic food packaging and waste by having 'nude food' days. You can also participate in activities such as the Aussie Backyard Bird Count. There are many different ways to make a difference and work with others towards a civic goal.

Greening Australia (www.greeningaustralia.org.au) engages Australians about environmental problems in a practical way. They are passionate about protecting and restoring the health, diversity and productivity of our unique landscapes. Greening Australia uses a combination of science and hands-on action with communities and volunteers to tackle key issues like climate change, biodiversity loss and conservation of native vegetation and wildlife habitat.

The **Rural Fire Service** (RFS) is known by a different name in each state, but their goal is the same: to protect the community from bushfires. The RFS trains volunteers and provides them with qualifications they can use in their everyday lives. They also provide all the personal protective clothing and equipment you need. In NSW there are over 2100 rural fire brigades with approximately 72 000 members. It is the world's largest volunteer fire service. These volunteers played a vital role in fighting the mega-fires across the state during December 2019 and January 2020.

Australian volunteer fire services:

- NSW Rural Fire Service (www.rfs.nsw.gov.au)
- Queensland Rural Fire Service (www.ruralfire.qld.gov.au)

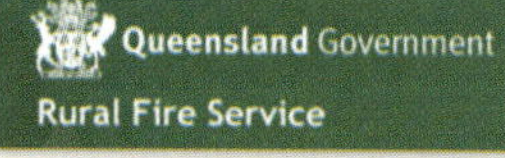

- Victorian Country Fire Authority (www.cfa.vic.gov.au)

- South Australian Country Fire Service (www.cfs.sa.gov.au)

- Western Australian Bush Fire Brigade (www.dfes.wa.gov.au)

- Tasmania Fire Service (www.fire.tas.gov.au)

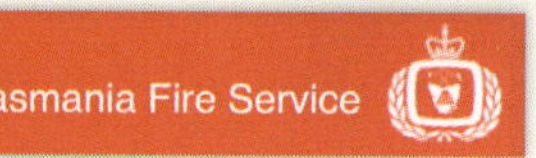

Source: Amazing Facts about Australia: *Wildlife Conservation*, pp. 69–71, Steve Parish

Research

There are lots of volunteer organisations all over Australia. Conduct your own research about one that isn't mentioned in the text. List five interesting things about the organisation.

1 ______________________________

____

2 ______________________________

____

3 ______________________________

____

4 ______________________________

____

5 ______________________________

____

Questioning

Imagine you could interview a volunteer wildlife carer. What would you ask? Write four questions.

1 ______________________________

____

____

2 ______________________________

____

____

3 ______________________________

____

____

4 ______________________________

____

____

Analysing

In recent years, the number of adults doing volunteer work has dropped from 34% to 31%. What are some reasons that could explain this?

____

____

____

____

____

____

Communicating

What issue is of special interest to you? Is it wildlife protection, beach safety or helping the elderly?

Prepare a 1–2 minute speech to encourage people to volunteer and support your special interest.

Evaluating and Reflecting

What are some useful strategies organisations could use to increase the number of volunteers?

____

____

What impact would increased volunteer rates have on the community?

____

____

____

How do I choose?

UNIT 29

Some chocolate icecream or a strawberry donut? A new gaming device or the latest sports shoes? Going to the movies or visiting the water park? Ahhhh – there are so many things that I want but I just can't choose!

Unfortunately, I don't actually need any of those things. **Needs** are items that are necessary for my survival, such as food, clean air and water, clothing, shelter and sleep. If I don't have what I need, then I won't be able to live and be healthy.

On the other hand, **wants** are things that I desire. They make life more fun and enjoyable, but I don't really need them. Personal wants are goods like bikes or computer games and services such as sports coaching or a haircut. Community wants – goods such as playground equipment and services such as street cleaning – are desired by many people in the same area.

Anyone who creates goods or services, whether it is for a want or a need, is called a **producer**. Anyone who pays to buy or use these goods and services is called a **consumer**.

The main problem with wants is that they are almost unlimited. We are never satisfied with the variety and quantity of the goods and services we consume. There is always something 'else' to have. This leads to the economic problem of **scarcity**.

Because consumers can't have everything that they want, they are forced to make choices between competing wants. For instance, due to my limited resources (how much money I have), I can buy either a hat or a pair of shoes. My decision will be based on my analysis of my personal circumstances at that point in time. Which is more important? How much will each cost? How many hats and pairs of shoes do I already have? Why do I want them?

Governments usually provide for the needs and wants of the community, and they face the same questions when making decisions about how to spend their limited resources. For example, if the government spends too much money on building playgrounds, there would be less money available to spend on upgrading hospitals.

But money is not the only factor in determining how consumers satisfy their needs and wants. The resources used in the production of goods and services are also limited. Once a tree is cut down to build a chair, it will take many years for another tree to grow to replace it. And it took millions of years for the Earth to create the coal we burn to give us electricity.

As a result, producers and consumers need to make careful choices when satisfying needs and wants to ensure the sustainable and efficient use of resources.

ECONOMICS AND BUSINESS

TARGETING HASS 5 © PASCAL PRESS ISBN: 9781925726060

Research

Interview at least five people to find out their three most important wants. Record the information.

Name: ______________________

Wants: ______________________

____

Name: ______________________

Wants: ______________________

____

Name: ______________________

Wants: ______________________

____

Name: ______________________

Wants: ______________________

____

Name: ______________________

Wants: ______________________

____

Questioning

Choose one person you interviewed for the Research activity. Write three questions that would help you find out more about their most important want.

Ask the questions and record their answers.

Question 1: ______________________

____

Answer 1: ______________________

____

Question 2: ______________________

____

Answer 2: ______________________

____

Question 3: ______________________

____

Answer 3: ______________________

____

Analysing

What did you notice about the wants that different people had? How were they similar or different? What reasons are there for these similarities and differences?

____

____

____

____

Communicating

Using the data gathered in the research activity, create a visual representation of this information to share with others. You could make a graph, a collage or a chart.

Share and compare your representation with others.

Evaluating and Reflecting

Would it be better if there were less choices available to consumers? For example, if everyone had the same shoes and there was only one type of bread to buy? Explain your reasons.

____

____

____

UNIT 30

Resources and Food Production

Food is a need. Without regular food consumption, we would not be able to survive. But producing food, whether it is on a farm or in your own backyard, uses resources such as sunlight, fertiliser and importantly – water.

Australia is a country known for its very dry climate. For many years, much of the country has been in drought. Communities regularly endure water restrictions that limit the use of water when washing cars, watering the lawn and even when taking a shower. But what many people don't realise is that a large percentage of their water usage is 'hidden' in the food they eat and the clothes they wear. By analysing the food production process, we can gain a greater appreciation of the use of Earth's precious natural resources.

Chocolate production

According to Australia's CSIRO (Commonwealth Scientific and Industrial Research Organisation), in 2015 the average Australian consumed 32 kg of chocolate. Since then, consumption levels have only increased. Now this might be alarming when you consider the health impacts, but when you look at the resources used, it's frightening.

To produce each 100 g bar of chocolate, 1700 litres of water are used. That is the equivalent of almost 10 bathtubs full!

Most of the water is used in growing the cacao tree in environments close to the equator. Cacao plants take 3–4 years before they grow fruit. The fruit is often harvested by hand, because the tree is delicate. The fruit pods are collected in baskets before being split open using a machete, freeing the beans. The outer husk of the pod is discarded. It takes about 1200 cacao beans to produce 1 kg of chocolate.

The beans are then covered with banana leaves and fermented for up to 8 days to help develop the chocolate flavour. During this time, the beans sweat off a pale-yellow liquid that is usually also discarded. At the end of the fermentation process, the beans are placed on wooden floors and dried in the sun for 1–2 weeks. The dried beans are then transported to a factory where they are sorted, cleaned and roasted. Machines crack open the roasted beans and blow away the shells, leaving the cocoa 'nibs' which are then ground into a paste (cocoa mass). The mass is pressed to squeeze out the cocoa butter.

The final step in the process is to blend the cocoa butter with other ingredients such as sugar, milk or milk powder and vanilla to produce different types of chocolate. Additional resources are used when packaging and transporting the chocolate to retail outlets for sale, and when using chocolate as an ingredient in other products.

TARGETING HASS 5 © PASCAL PRESS ISBN: 9781925726060

Research

What are some of your favourite foods?
Can you find out how much water is used in their production?

Food	Water used / unit

Questioning

What were some questions you thought of while reading this text? Write three questions.

1 ______

2 ______

3 ______

Analysing

Apart from the amount of water used, in what other ways could chocolate production appear to be wasteful?

Communicating

Create a flowchart using images and text captions to illustrate the production process of chocolate, or another food product.

Evaluating and Reflecting

Does the information in the text make you reconsider how much chocolate you eat? Explain your answer.

Use of Resources

Indigenous Australian peoples, like many indigenous cultures around the world, traditionally relied upon the environment to provide everything they needed to live and thrive. For this reason, they learned to sustainably use the resources available. This means they managed their environment in ways that kept it healthy so that it could provide for them for generations to come. This is how they achieved it.

Moved with the seasons
Some Indigenous Australian groups moved seasonally from area to area, as the amount of food varied throughout the year. This provided them with a continuous supply of food and enabled food sources to re-establish before being harvested again.

Used a wide variety of plants and animals
They understood that all forms of nature, including humans, depend upon each other for survival. They used a wide variety of plants and animals to ensure the supply of any one plant or animal never ran out, avoiding any knock-on effects. They took only what they needed from the land and never wasted anything.

Lived by the patterns of nature
Over thousands of years, Indigenous Australian peoples observed the land and its animals and plants. They became familiar with the seasons and knew that with each season came change. Some seasons were better for hunting or fishing, while others were better for collecting fruits and vegetables, or for preparing medicines. They planned their lives around the cycles of nature.

Fire-stick farming
They used fire in a sustainable manner to make the environment more suitable for humans. They did this by regularly burning small patches over large areas. This limited the intensity of large bushfires as there was less to burn. In addition, the regrowth after the fire attracted animals, which could be hunted as they fed off the fresh green shoots. The burning created a pasture-like landscape, which made it easier to move and hunt.

Shared knowledge
They passed down the knowledge of how to live a sustainable life, and the importance of doing so, to the younger generations through stories, songs, art and dance.

Source: Australian Geography Centres: *Upper Primary*, p. 51, Blake Education

ECONOMICS AND BUSINESS

TARGETING HASS 5 © PASCAL PRESS ISBN: 9781925726060

Research

What is the difference between natural resources and capital resources? List three examples of each type in your answer.

Questioning

For each answer, write a question that is based on the text.

Answer: regrowth
Question:

Answer: natural cycles
Question:

Answer: interdependence
Question:

Answer: seasonal
Question:

Analysing

What natural resources are referred to in the text, and how were they used sustainably by Aboriginal peoples?

What capital resources are referred to in the text, and how were they used sustainably by Aboriginal peoples?

Communicating

Create a seasonal chart to show what fruits and vegetables are harvested each season in your state or territory.

Evaluating and Reflecting

Can people today learn from the methods Indigenous Australians used to manage the environment? What would be the most important lesson? How could we change our modern behaviours?

UNIT 32

Following Fads

WHAT IS A FAD?

When someone refers to a 'fad', they usually mean something that a group of people gets very excited about for a short amount of time. This can be an interest, a toy, or even a type of dance or fashion. Some fads become popular again after a long time as a new generation discovers them. People often look back at fads fondly when they think about the fun they had at the time, even though they may not be excited about the fad any more.

WHEELED FADS

Devices with wheels have been a favourite with kids of all ages for a very long time. Back in the 1950s and 1960s, children often built their own go-karts to race each other. In the 1970s, people enjoyed rollerskating at roller discos, where they played disco music while people skated around a rink. In the 1980s and 1990s, colourful inline skates known as Rollerblades became popular. Today, it's very common to see kids riding down the street on the latest scooters and skateboards. Rollerskates have made a comeback in the past few years, as they are used in the competitive sport known as roller derby.

Ask some adults what fads they experienced while they were growing up. Are any of these fads popular again?

TOY FADS

Over the years, there have been many toy fads. During the 1940s, people enjoyed simple toys like the Slinky and the Magic 8 Ball. In the 1950s, kids got active with the Hula-Hoop and the Frisbee. The 1960s saw the introduction of the Etch A Sketch and radio-controlled cars, while the 1970s were dominated by NERF and Star Wars action figures. It seemed like almost every toy had a television series in the 1980s, be it Care Bears, My Little Pony or Transformers. In the 1990s, toys moved with technology — you could feed your Tamagotchi and even talk to your Furby.

WEIRD FADS

There have also been some strange fads. In the 1920s, flagpole sitting was a popular spectator 'sport' for stunt performers and publicity seekers. During the Great Depression of the 1930s, dance marathons became a popular distraction as people danced almost non-stop for hundreds of hours. One marathon even lasted for more than 16 weeks!

What fads are being enjoyed now? Have you ever been caught up in one? Write a list with an explanation of each fad and whether you think it will last long or not.

Source: *Blake's Australian History Guide*, p. 106, Pascal Press

TARGETING HASS 5 © PASCAL PRESS ISBN: 9781925726060

Research

Find one product (a good, not a service) that has been a fad during the past five years.

What was the name of the product?

___

What did the product do?

___

Why was it so popular?

___

When did the fad start and how long did it last?

___

How much did the product cost? ___

Who was the target market of the product? That is, who was it aimed at?

___

Questioning

True or false?

1 Fads always use wheels in their products.

___

2 Many fads are aimed at children.

___

3 Fads never return once they are finished.

___

4 Toy fads are often linked to television programs and movies.

___

5 Social media is not involved in creating fads.

___

6 Anyone can start a fad. ___

Analysing

Using the fad from your research, identify the factors that influenced customers to purchase it. Tick the relevant factors and give a brief explanation.

	Factor influencing consumers	Explanation of its use in marketing the product
	Budgeting (low cost)	
	Social media	
	Celebrity endorsement	
	Cultural relevance	
	Environmental benefit	
	Ethically desirable	
	Peer pressure	

Communicating

Discuss with a friend a fad that you have been involved in.

What was it?

Did you make a purchase?

What influenced you to take this action?

How did you pay for the product?

Do you still use it?

Where is it now?

Evaluating and Reflecting

Looking back at your discussion with a friend, was your experience with the fad positive or negative? Explain your answer. How was your friend's fad experience similar and different to yours, and why?

___

___

___

___

History Assessment 1

ASSESSMENT

What do we know about the lives of people in Australia's colonial past and how do we know?

Tasks

- Analyse these two images from colonial Australia.

Gold was discovered near Coolgardie, Western Australia in 1898.

A cartoon entitled *Back Country Squatter*, AD 1892

What type of source material is it?

Is it a primary or secondary source?

What is the purpose of the image?

What do you observe about the image?

What does the image tell you about the lives of people in Australia's colonial past?

What feeling do you think the image is trying to convey, and why?

What additional information would you need to know to have a better understanding of the ideas explored in this image?

What type of source material is it?

Is it a primary or secondary source?

What is the purpose of the image?

What do you observe about the image?

What does the image tell you about the lives of people in Australia's colonial past?

What feeling do you think the image is trying to convey, and why?

What additional information would you need to know to have a better understanding of the ideas explored in this image?

Sources: Go Facts Australia: *Colonies*, p. 11, Blake Education; Amazing Facts about Australia: *Early Settlers*, p. 27, Steve Parish

TARGETING HASS 5 © PASCAL PRESS ISBN: 9781925726060

History Assessment 2

What were the significant events and who were the significant people that shaped Australia's colonies?

Tasks

Look at the images of groups of people who influenced the development of the colony. For each group:

- outline their role in colonial development
- give some examples of related historical figures and events.

Overland explorers

Bushrangers

Native police

Chinese goldminers

Sources (from top to bottom): Amazing Facts about Australia: *Early Settlers*, p. 68, Steve Parish; *Blake's Australian History Guide*, p. 66, Pascal Press; *Blake's Australian History Guide*, p. 65, Pascal Press; Australian History Centres: *Upper Primary*, p. 27, Blake Education.

History Assessment 3

How did an Australian colony develop over time and why?

Between 1788 and 1900, the colony of New South Wales grew from a small penal settlement to a society on the edge of becoming a nation. What significant events occurred to make this happen? How did the colony develop?

Task

- Create a timeline of the colony, showing its expansion as well as identifying the impact of significant events and people. You can include maps, drawings and diagrams.

TARGETING HASS 5 © PASCAL PRESS ISBN: 9781925726060

Geography Assessment 1

How do people and environments influence one another?

Task

- Using the aerial photographs, answer the following questions in a way that explores the influence people and environments have on each other.

Fairfield, Queensland. Before and after the flood of 2011.

https://www.abc.net.au/news/specials/qld-floods. First posted 3 June 2011 at 12:33 pm

1 How does the local environment impact on the people who live in this area?
Hint: Think about influences on people's lifestyles and use of spaces before as well as after the flood.

____

____

2 What changes have people made to the natural environment that may have affected the severity of the floods in this suburban landscape?

____

____

3 As a result of the floods, what changes to the local environment could be made by people and why should they make these changes?

____

____

____

Geography Assessment 2

How do people influence the human characteristics of places and the management of spaces within them?

ASSESSMENT

Task

- Read the information and answer the following questions.

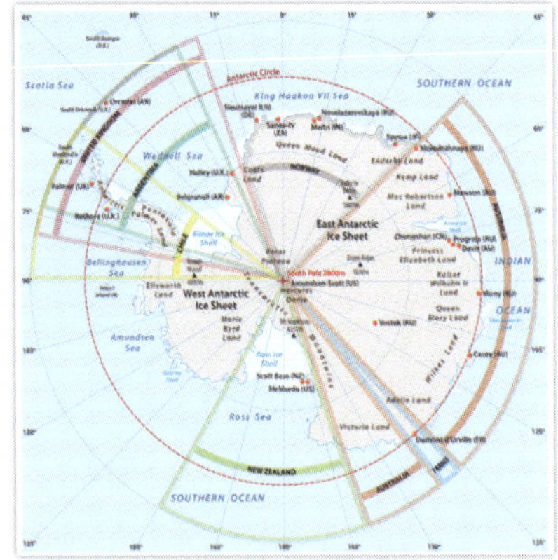

Antarctica is located in the southern hemisphere. It is the coldest, windiest and driest continent in the world. In winter and autumn the sun never rises, and in summer and spring it never sets.

Many scientists visit Antarctica each year to study the effects of climate change and Antarctic wildlife and fisheries.

Seven nations claim territory in Antarctica: Australia, Argentina, Chile, France, New Zealand, Norway and the United Kingdom. About 70% of the world's fresh water is contained in the ice and snow of Antarctica.

Humans have left an environmental footprint on Antarctica. This means that, just like a footprint, they have left a mark on the environment. They have produced sewerage, waste and air pollution; spilled fuels such as oil and petrol; and abandoned buildings. They have even accidentally brought in non-native seeds, mould spores and small animal eggs via their clothes, shoes and equipment. In the past, humans have also hunted whales and seals to the verge of extinction. Human waste is a potential threat to Antarctica and must be managed carefully. Most waste produced at Australian research stations is transported to Australia for disposal.

The Antarctic Treaty was made in 1959. It is an agreement between several countries about how to divide the area up, how to manage it, and how to protect it. Countries that have signed have agreed to use the land for peaceful purposes only, ban nuclear weapons and waste from the area, and protect the environment.

1 What aspects of Antarctica make its environment difficult to manage?

____

____

2 How have people's actions negatively impacted the Antarctic region?

____

____

____

____

____

3 Have there been any ways to solve or ease these negative impacts?

____

____

____

____

____

4 What would be the possible effects of poor management of the Antarctic environment?

____

____

____

TARGETING HASS 5 © PASCAL PRESS ISBN: 9781925726060

Geography Assessment 3

How can the impact of bushfires or floods on people and places be reduced?

1 In the past 10 years, which states have seen the worst bushfires? ____________________

2 What climactic conditions result in an increased risk of bushfires and their severity?

____

3 How do bushfires impact the natural environment in a positive way?

____

____

4 How do bushfires impact the natural environment in a negative way?

____

____

5 Natural events such as bushfires can have a devastating impact on people and communities, as well as environments. However, if we are prepared, we can minimise some of the negative consequences. Create a flow chart for the development and implementation of a bushfire prevention plan.

My Bushfire Action Plan

	Action	Reason
Step 1		
Step 2		
Step 3		
Step 4		
Step 5		

Civics and Citizenship Assessment

What is democracy in Australia and why is voting in a democracy important? How and why do people participate in groups to achieve shared goals?

Task

- Read the information and complete the activities.

It is a new school year and your school principal has created a new set of rules for your school.

THE NEW SCHOOL RULES

Rule 1
Students must not speak unless asked a direct question by a teacher.

Rule 2
Students can only play and work with peers chosen for them by their teacher.

Rule 3
Teachers can eat student lunches whenever they like. They do not have to offer any compensation.

Rule 4
Students whose first name is only one syllable must attend school on the weekends as well as weekdays.

Rule 5
Teachers can put students on detention at any time, for any reason, without explanation.

1 Do you agree with these new rules? Why or why not?

____

2 How do these rules compare with Australia's democratic values?

____

____

3 What action would you take if these rules were really implemented in your school?

____

4 Why would this action be a good idea?

____

____

5 If you became school principal, what three rules would you create?

____

____

____

6 How would you make sure that your rules met Australia's democratic values?

____

____

Our freedoms in Australia are often seen as our 'rights'. But with these rights also come responsibilities. For example, we all have the right to express an opinion (free speech). Our matching responsibility is to speak without malice and hatred.

7 Take the three rules you would like to introduce if you became school principal. What responsibilities for teachers and students accompany those rules?

Rule 1: ______________________________

Rule 2: ______________________________

Rule 3: ______________________________

TARGETING HASS 5 © PASCAL PRESS ISBN: 9781925726060

Economics and Business Assessment

Why do I have to make choices as a consumer? What can I do to make informed decisions? What influences the decisions I make?

Task

- Use the advertisement to answer the following questions.

1 Is a visit to the Little Boring Day Spa a **good** or a **service**? Why?

____

____

____

____

2 Is a visit to the Little Boring Day Spa a **need** or a **want**? Why?

____

____

____

____

3 What types of resources would be used in the day spa? List two for each category.

Human resources	Natural resources	Capital resources

4 What strategies does this advertisement use to persuade you to spend your money at the day spa?

____

____

5 Are these strategies effective in influencing your choices as a consumer? Explain why.

____

____

6 How would the issue of 'scarcity' influence your decision to spend money on a visit to the Little Boring Day Spa?

____

____

____

Answers

Note: Answers are not supplied to open-ended questions, where students' responses will vary.

1 Past Lives

Research: Settlers had to bring agricultural tools such as ploughs, hoes and spades; and implements like hammers, nails and axes which were in short supply in the colony. They had to supply their own crockery, cutlery and blankets. They also needed to bring food supplies such as flour and tea. These supplies were heavy and bulky and transporting them to their plot of land was hard.

Questioning: The primary historical sources are the painting of the squatter's hut, the ruins of the stone cottages and the artefacts found in the huts. I know this because they are sources that existed or were created at the time, not later.

The secondary source is the illustration of the hut. This is because it is a source created by someone who did not experience the event firsthand or participate in the events.

Analysing: Yes because it was painted during the 1800s and is likely to be accurate in its detail. The items in the painting were real items used in the 1800s.

2 Convict Life

Research:

Norfolk Island – The first penal settlement was established to stop the French from colonising it. Fifteen convicts and 7 free men landed on 6 March 1788. Their settlement of the island failed when crops perished due to poor weather conditions and rats. The timber they were going to cut down for ship's masts was also not suitable. When the colony in Sydney was facing starvation, many more soldiers and convicts were sent to Norfolk Island. In 1803 the penal colony was abandoned because it was too remote and cost too much to maintain. However, in 1824 the British government re-established the penal colony for the 'worst' of the convicts, men who had been 'twice convicted'. It was a place of extreme punishment and hardship: food was scarce, housing was inadequate, and convicts were tortured and flogged. In 1847 the British government started winding down the penal settlement and in 1855 the last convict on the island was taken to Tasmania.

Port Arthur – Between 1833 and 1853, 73 000 convicts were transported to the penal colony at Port Arthur in Van Diemen's Land, making it the main penal colony in Australia. Its convicts were secondary offenders, or those who were rebellious. Most of the male convicts at Port Arthur were sentenced to manual labour because the government wanted the colony to be self-sustaining. So they cut timber, worked in gardens, made shoes and even helped in the shipyards. Prisoners at Port Arthur were rewarded with food (such as tea and sugar) for being well behaved, and had food taken away if they made trouble. Other methods of punishment used in Port Arthur were the cat-of-nine-tails, the wearing of leg irons and solitary confinement. Prisoners were up by 5am for Bible reading and prayers and they worked until 5pm. Once convicts had obtained their ticket of leave, many left the island and settled in Victoria.

Questioning:

1. 2.2 pounds in one kilogram
2. A ticket of leave is a document giving freedom (a pardon) to a convict who has served their sentence.
3. You would be given the cat-o'-nine-tails as a punishment.
4. If you were given a pardon, you were forgiven for the offence and you would not have to serve the rest of your sentence.

Analysing: The artefacts originated from the penal settlement of Van Diemen's Land. They accompany the text to provide additional information to the reader to help them understand historical terms such as chain gang. The impact on the reader is to help them develop a greater understanding of the daily life of the convicts.

Communicating: Responses will vary, but may include:

Term (word)	Definition in your own words	Picure or illustration
barracks	A long rectangular hut or room lined with bunk beds where convicts slept	Will vary
ration	The amount and type of food given to all convicts each day	Will vary
stocks	A wooden structure used for punishing convicts	Will vary
cat-o'-nine-tails	A whip with nine strips used for punishing convicts	Will vary
ticket of leave	A document allowing convicts to work for themselves	Will vary
chain gang	A group of convicts chained together, working on public projects such as building roads	Will vary

3 Growth of Australian Colonies: 1800s

Research: Answers will vary, but may include:

- From 1967 to 1992, Caroline Chisholm's portrait was on the Australian $5 note.
- She was born in Northhampton England in 1808.
- She came to Sydney in 1838.
- She helped bounty women find jobs and she let many of them stay in her home.
- She taught them basic cooking skills.
- She returned to England in 1846 but kept helping people emigrate to Australia, reuniting them with family.
- She returned to Australia in 1854, working to improve conditions in the goldfields.
- Caroline retired to England in 1866 and died in 1877, poor and mostly forgotten.

 TARGETING HASS 5 © PASCAL PRESS ISBN: 9781925726060

Answers

Note: Answers are not supplied to open-ended questions, where students' responses will vary.

Questioning: Answers will vary, but may include:

- What skills do you have?
- Do you have any other family members or are you married?
- Are you in good health?
- Why do you want to travel to Australia?

Analysing:

People migrated to Australia for many reasons: to seek a better life, to find a husband, to get a job, to escape the potato famine, to farm on their own land.

The type of person most likely to migrate would have been someone who was negatively impacted by the industrial revolution, the potato famine or the Scottish Highland clearances. This is because they had more to gain by migrating and taking the risk of starting again in the colony.

4 Queensland 1859

Research: Answers may include:

The Old Windmill in Brisbane is heritage listed. It is also known as Brisbane Observatory. It is the oldest surviving convict-built building in Queensland. In 1992, it was put on the Queensland Heritage Register. It is in Observatory Park. It used two pairs of millstones to grind wheat to make flour. When there was no wind, convicts would make it work by walking on a treadmill. This was a form of punishment.

Other convict-built structures were the Colonial stores building and Commissariat store, the immigration depot which became the National Trust House and Newstead House.

Questioning: Answers will vary but may include questions about tools and equipment used, how long it took, how the convicts were treated and how they lived.

Analysing: Sources like the watercolour paintings provide information about types of housing and building materials. They tell us about where structures were in relation to each other, and about the use of technology at the time for transport and communications. They can also tell us about the types of clothing worn by different groups of people and how those groups interacted.

5 Colonisation and its Impact on Indigenous Peoples

Research: 1788 (January) – colony founded in Sydney. 1788 – (November) Parramatta settled. 1791 – Windsor settled in western Sydney. 1794 – Richmond. 1798 – Liverpool. 1790s – expansion into the Hunter and Illawarra regions. 1804 –Newcastle. 1803 – Blue Mountains were crossed, opening up the western regions of the state. 1815 – road built over Blue Mountains. 1817 – Shellharbour was settled. 1818 – Penrith became a town. 1820 – settlement built at Bathurst.

Questioning:

1. The Aboriginal peoples did not have any immunity to these diseases because they had never been exposed to them before.

2. The government controlled the wages so that they could control the Aboriginal people and keep them from moving too much.

3. Colonials thought that Aboriginal people weren't like them. They thought Aboriginal people needed to be protected and that their cultures and beliefs should be changed to be the same as the Europeans'.

Analysing: Colonisation meant that Aboriginal people were separated from Country. They were not able to hunt and live in their traditional ways. They had to wear European-style clothing and learn English. They had to stay in the same place and live in European-style housing. They had to work as farmers and labourers. They began to drink alcohol.

Evaluation and Reflection: Answers will vary, but may include reference to the Aboriginal peoples' knowledge of the land, its animals, plant life and seasons. They knew what could be eaten and what was poisonous.

6 Land Clearing and Introduced Species

Research: Five rabbits were brought over with the First Fleet, but they were not released into the wild. In 1859, 13 rabbits were released in Victoria. There was lots of food and few predators so they multiplied rapidly. By 1880, they had crossed into New South Wales. By 1886, they were in Queensland and by 1894, in Western Australia. They ate food meant for grazing animals and native mammals. Their burrows caused erosion and the loss of farmland. The government tried to control their population by trapping and offering a bounty for their skins. A 3256 kilometre rabbit-proof fence was built across Western Australia, but it was too late. By the late 1940s there were approximately 600 million rabbits in Australia.

Questioning:

acclimatisation: the process in which an animal or plant adjusts to a change in its environment

cochineal: a red coloured dye used for food colouring

resilient: able to recover quickly from difficult conditions or events

export: sending goods or products to another country to be sold

infested: to exist in large numbers, causing damage

Analysing: Answers will vary, but may include reference to the loss or damage of natural features which have spiritual and cultural significance. It would also negatively impact the animals and plants that Aboriginal people traditionally relied on for food and as their totems connecting them to Country.

7 Governor Macquarie

Research: Hyde Park, the Domain, the Botanical Gardens, the grounds of Sydney University, the area that is now Centennial Park

Answers

Note: Answers are not supplied to open-ended questions, where students' responses will vary.

Questioning: Answers will vary, but may include questions about how many people will be using it, how it will be accessed by people, who will pay for its development, and how it's construction can be done sustainably.

Analysing: Answers may include the pollution of natural water sources, destruction of native animal habitats, disturbing Aboriginal sacred sites and impacting their connection to Country, spreading disease, dispossessing Aboriginal peoples of their land.

8 Eureka Stockade

Research: A royal commission recommended that the licensing laws be replaced. Miners did not have to carry their licence with them at all times. Instead, they paid a tax on the gold they found. Miners were also given the right to own the land they were mining. The commission also recommended that Chinese immigration be restricted. Police numbers were cut and the gold commissioners were replaced by mining wardens. The state parliament's legislative council was expanded to allow for greater representation.

Questioning: Answers will vary, but may include questions about if the wife fears for her husband's safety, if she helps him in the mining process, what their everyday life is like, what she cooks for meals.

Analysing: The flag symbolised the miners' protest against government authority. It was a symbol of their desire for liberty. The white cross joining the stars symbolises their unity. To the government, the flag symbolised the miner's defiance and an attack against their own power and control of the people and the goldfields.

9 Myall Creek Massacre

Research: One Tree Hill, in southern Queensland, was a series of conflicts in 1843 between European squatters and Aboriginal people. Several of the local Aboriginal tribes were in an alliance, led by Multuggerah. Multuggerah and his warriors ambushed a supply convoy of bullock drays by building barricades across the track. The Europeans fled, but regrouped and attacked the Aboriginals who had camped nearby. Several Aboriginals were injured and they appeared to retreat higher up on Mt Tabletop (One Tree Hill). They had stacked spears and boulders there, and the attacking Europeans were surprised and once again defeated. Finally the border police arrived, but their leader decided they did not have enough men to win a battle with the Aboriginal warriors at the time. It was one of very few victories by Aboriginal people over European settlers.

The similarity between One Tree Hill and Myall Creek is that they were both conducted by Aboriginal people from different tribal groups working together. The differences: Myall Creek massacre involved children and there was a trial after the event.

Questioning: Answers will vary, but may include questions about the facial expressions on the different groups of characters; what is shown by the European men riding horseback and the Aboriginal people walking; and how the artist shows the lack of power and control Aboriginal people have in this situation.

Analysing: It shows the reader that there were colonials who were sympathetic to the Aboriginal's situation, "slaughtered without having given the slightest provocation". It is a primary source as it was written at the time by a person observing the community's reaction. It isn't completely reliable as it is based on one person's opinion. It tells you that the community was upset by the events.

10 The Chinese in Australia

Research: The Lamming Flat riots occurred in 1860–61 on the goldfields of NSW. The European miners resented the way the Chinese miners worked in large organised groups. The Chinese miners lived very simply and could live on much less money than the Europeans. The Europeans claimed that the Chinese muddied the water holes, were thieves and worked on weekends. This, combined with their different appearance and language, made them targets of racism. After months of tension, more than 2000 European miners attacked the Chinese miners and forced them off the goldfields at Lamming Flats. Tents were destroyed and the Chinese miners' possessions were taken. There were many beatings but no-one was killed.

Questioning:

1. Coolies were employed because they were cheap labour.
2. Chinese migration led to the development of market farms as they grew fruit and vegetables on the goldfields.
3. The economy grew as a result of Chinese migration: the population increased and so did the number of small businesses and restaurants.
4. The Chinese population became concentrated in cities due to government immigration restrictions.

Analysing: Answers will vary, but may include reference to language barriers, lack of money, no friends or family to help, facing racism and discrimination from existing miners.

Evaluating and reflecting: Answers will vary, but should include impacts on the food that we eat, the way food is bought and sold, and Australia's willingness to be a multicultural society.

11 Afghan Cameleers

Research: Sample answers include:

Burke and Wills – 26 camels. They explored from Victoria, up through central NSW and to the southern end of the Gulf of Carpentaria.

Peter Warburton – 17 camels. Explored from Adelaide to Alice Springs and then west to the Oakover River north of Perth.

TARGETING HASS 5 © PASCAL PRESS ISBN: 9781925726060

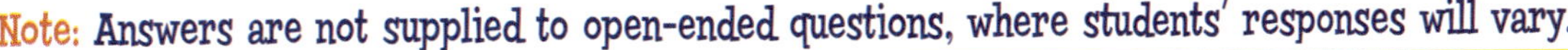

Note: Answers are not supplied to open-ended questions, where students' responses will vary.

Ernest Giles – number of camels unknown. He made five different explorations from South Australia, heading west into Perth. He explored past The Olgas and through the Gibson Desert.

David Lindsay – 44 camels. He explored western South Australia and into Western Australia to the head of the Murchison River.

Edward Kidson – 12 camels. He explored the Great Sandy Desert.

Cecil Madigan – 19 camels. He explored the Simpson Desert from Andado Station in the Northern Territory to Birdsville.

Questioning: Sample answers include:

Why did the Afghans come to Australia?

Why are camels no longer used as a method of transport?

Where did the cameleers come from?

How were the cameleers discriminated against?

Where can aspects of Afghan culture still be found in Australia today?

12 Pioneering Women

Analysing: The images tell you about the clothes they wore, the types of houses they lived in and what the houses were made of, the tools and implements they used, and the types of jobs they did every day. The images support the text in their impression of the hard life faced by pioneering women.

Communicating: The qualities include: strength, resilience, persistence, innovative, committed, self-reliant. They are pioneers because they prepared the way for others to follow. Oral presentations will vary.

Evaluating and reflecting: Answers will vary, but may include references to slower development of the colony; slower levels of expansion; less innovation.

13 Europe

Questioning: Sample answers include:

hemisphere: What is the half of the Earth north or south of the equator called?

Eurovision: What is the singing contest for European countries called?

Vatican City: Which city is a tiny country inside Rome?

Mt Elbrus: What is the highest mountain in Russia?

Analysing: Sample answers include similarities in language and culture, historical relationships between regions and people, and the ease of transporting goods between countries. Also, they are all in the same continent and many now share the Euro as a common currency.

Evaluating and reflecting: Answers will vary, but may include references to the fact that landmarks are now more crowded, they have more visitors every year, maintaining/ cleaning them is more difficult with increased human traffic.

14 The Netherlands and Las Vegas

Research: Much of the Netherlands is an alluvial plain and 2000 years ago most of the country was covered in peat swamps. There was very little higher ground on which to build houses and settlements. As the population grew, there was an increasing need for land. As the sea tides rose and fell, they left behind layers of sediment which slowly built up. The Dutch people used these to extend the existing dykes and move them further outward towards the sea.

Questioning:

aquifers: underground layers of porous rock that hold water which can be extracted

irrigation: the delivery of controlled amounts of water to plants at specific intervals

reclaimed: to take land that was waste or once underwater and return it to being usable by humans

canal: natural or artificial waterways or channels that carry water, often to help with irrigation or transportation

dam: a barrier that stops or slows the flow of water

Analysing: Possible answers include:

Positive aspects of growth – more money in the economy, increased jobs, increases in taxes taken by the government which can then be spent on schools and hospitals.

Negative aspects of growth – damage to the natural environment, use of limited water resources, increases in unemployment, deforestation, increase in numbers of feral animals.

Communicating:

1. A windmill is used to harness the power of the wind. In the past, they were used to grind wheat into flour. Today, they are used to generate electricity.
2. It has a large flat disc which rotates.
3. Attached to the disc are a number of long arms called 'vanes'. These vanes are angled to help them catch the wind.
4. The disc is connected to an axel which is connected to a tall shaft.
5. At the bottom of the shaft is a grinding wheel, generator or drive belt, depending on the purpose of the windmill.
6. When the wind blows, the vanes are pushed.
7. This causes the disc to rotate, which in turn spins the shaft and drives the grinding wheel or generator.

15 The Trans-Alaska Pipeline

Research: Sample answers: Inuit are the indigenous people of the Arctic region. Traditionally, they wore parkas and mittens made from caribou or sealskin. They ate fish, seal, geese, ducks and reindeer, with very little fruit or vegetables, although berries and plants were gathered in the spring and summer months. When hunting, they would live in temporary igloos, huts made from ice and snow. Their permanent houses were made from turf or earth and large stones. Inuit would travel across the land using

Answers

Note: Answers are not supplied to open-ended questions, where students' responses will vary.

ANSWERS

a sledge, often pulled by dogs, and they used kayaks with double-bladed paddles made from whalebone or driftwood when travelling across the water.

Analysing: Answers may include reference to deforestation around the pipeline, pollution caused by the construction process and the trucks/machinery used, the destruction of native animal habitats and breeding areas, changes to the natural watercourses during snow melts.

16 American Dust Bowl

Research: The areas affected were in New South Wales around the towns of Dubbo, Broken Hill, Nyngan, Narromine and Parkes; in South Australia around Peterborough and Yunta as well as the Riverland region and Flinders ranges; and in Victoria where the dust made its way to Melbourne.

The wider impact of the dust storms included stripping the farming and grazing land of the nutrient-rich topsoil; polluting the air with small particles which negatively impact people's breathing, especially those with asthma; closing swimming pools because of the pollution caused by the dust settling in the water.

Analysing: The images tell you about the clothing that people wore, the modes of transport they used, the possessions they thought were important, and the types of houses they lived in. They show the devastating impact of the dust storms and how severe they were – enough to almost cover cars. Roads were hidden and farmland was destroyed along with infrastructure such as dams, fencing and sheds. People had to leave their homes and find shelter and jobs elsewhere.

Communicating:

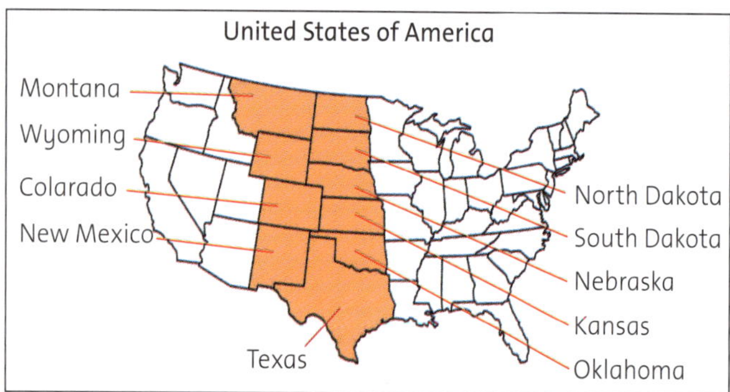

Evaluating and reflecting: Answers may include references to planting more trees and shrubs with wide-spreading root systems to hold the soil together; planting windbreaks on farms; using native plants rather than introduced species; building structures with better air filtration.

17 Responding to Climate

Research: Opals are a form of silica which contain water. They are classed as gemstones, not minerals. They are multicoloured. Black and red opals are the rarest and most valuable. Opals are found in sedimentary sandstone and mudstone which has been weathered, releasing the silica. Small cracks in the rock trap the water carrying the silica where it eventually hardens. Lightning Ridge in Australia is famous for its opal mining. Opalised plant and animal fossils have been discovered there.

Questioning: Answers may include questions about what it is like living in a house with no windows, how they manage the heat during the day, and what they do for sport and leisure.

Analysing: Coober Pedy: live underground, work underground in mines, little access to public transport, few hospitals and other public services, inside recreational activities, no parks or outside playgrounds

In Common: houses, attend church, go shopping

Perth: good access to public transport, variety of schools and hospitals, indoor and outdoor recreational activities, swimming pools and beaches, use air conditioning to control temperature inside, outdoor parks and play equipment.

Communicating:

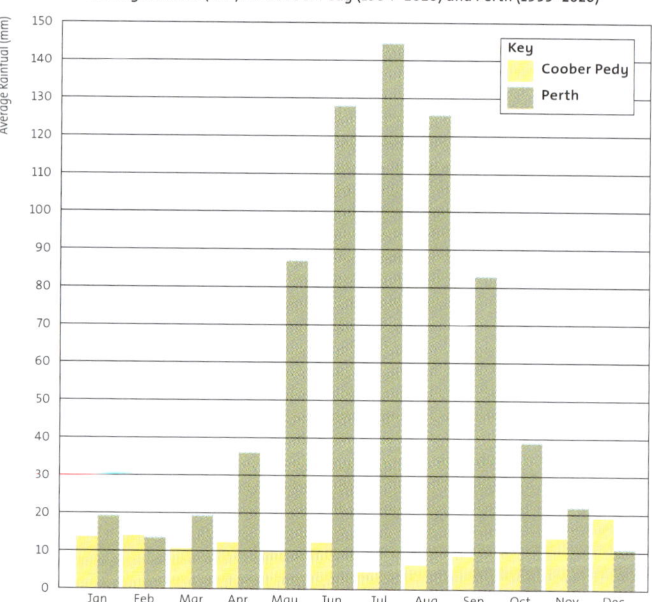

Evaluating and reflecting: Sample response: Most people prefer to live in places closer to the coast for the weather. It is not as hot, has more rainfall and so is a more pleasant living environment. Because of the better weather conditions, there are more recreational activities to participate in. Coastal cities also have a better range of health and education facilities, so people with young children and families usually prefer to live there.

TARGETING HASS 5 © PASCAL PRESS ISBN: 9781925726060

Answers

Note: **Answers are not supplied to open-ended questions, where students' responses will vary.**

18 Wooleen Station – Case Study

Research: Royds brought in time-controlled rotational grazing so that the land had time to rest and regenerate. They now use organic waste as fertiliser and stopped using chemical herbicides. They limited the number of animals they were grazing. They replaced ploughing the land with surface cultivation and direct sowing. They diversified the crops grown so there was more variety. They planted more native trees.

Questioning: Sample questions include:

How has the cover of native plants increased? How many introduced weeds species still exist on the property? How much topsoil is there now? Which species of native animals are found on the property and how many are there?

Analysing: There is greater ground coverage by small plants and shrubs, more places for native animals to shelter and find food, and less chance of soil erosion due to wind and rain.

19 Sustaining the Land

Research: The Indigenous rangers conduct controlled burns to reduce the build-up of fuel. They monitor the fires using satellite images and maps. They also create firebreaks which help prevent uncontrollable fires later in the dry season. They control introduced plant species and weeds, along with feral animals. They replant native trees and shrubs.

Questioning:

fallow: farmland that has been ploughed but is left for some time without being sown with seeds

germinate: when seeds begin to grow and put out small shoots

biodiversity: the variety of plant and animal life in a particular habitat

firebreak: an obstacle or barrier to prevent the spread of fire

sustainable: able to be maintained at a certain rate for a long time

Analysing: Traditional land management practices are sustainable because they work with the land and the seasons. Everything is used and nothing is wasted. They only use what is required, nothing extra. The land has been managed in this way for tens of thousands of years. If the land wasn't managed in this way, it is likely that populations would decline and die out.

20 Indigenous Aquaculture

Research: Some fish traps were created by rocks and bends in the river. Others were made by using rocks to create shallow pools which had openings that could quickly be blocked off. Some groups wove natural materials such as long grasses into cylindrical traps. Fish would swim into the trap where the space in it became narrower and they were stuck.

Questioning:

Aquaculture means the farming or harvesting of fish and other marine life.

The fish traps were different to adapt to local environmental conditions such as tides and water levels as well as the local species of fish.

The Brewarrina fish traps were sustainable because they allowed fish through, so not too many were taken at a time and the fish could reproduce.

The fish traps show the Aboriginal peoples spiritual connection to Country through their sustainability and the way that they used natural resources to take what they needed without waste.

Communicating: Sample answer:

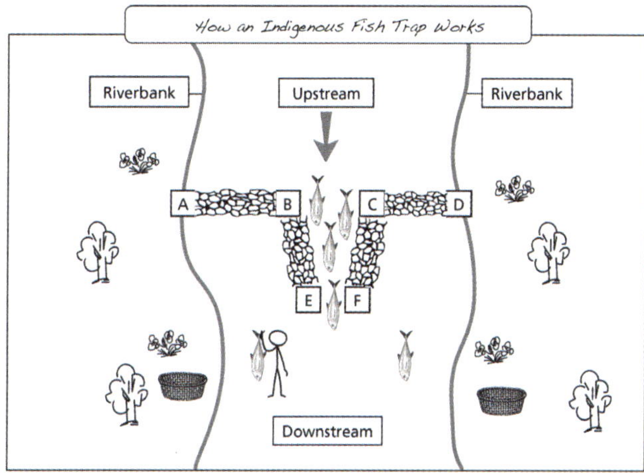

Evaluating and reflecting: Answers will vary, but may include comments about only taking fish they will eat themselves; not leaving fishing gear such as lines and hooks lying around as rubbish; and using natural materials in their fishing tackle instead of plastic lures.

21 Town Planning

Research: A wildlife corridor is an area of habitat, usually made up of native vegetation, which connects two or more larger areas of similar habitat that are separated by human activities or structures. They support native wildlife populations by allowing them to move to find food, water and mating opportunities. Answers for the second part of the question will vary.

Questioning: Sample answers:

industrial: In what zone would you find manufacturing businesses and transport hubs?

zone: What are different areas with specific features known as?

hospital: Which community feature provides for medical assistance?

cost: What is the financial burden of providing spaces for community use?

bandicoot: What is the name of an endangered Australian native animal that has been negatively impacted by human development in southern Victoria?

Answers

Note: Answers are not supplied to open-ended questions, where students' responses will vary.

Analysing: The table provides information about the types of zones, what spaces are found in these zones and a brief description of each zone's specific features. It is useful to show the similarities and differences between zones and how they are used by people, as well as how they may impact the environment.

22 Bushfires and Floods

Research: Lake Eyre sits in a natural drainage basin, so many rivers from Queensland, South Australia and the Northern Territory flow into it. When it rains in Queensland, the major rivers there, the Torrens, Thomson, Barcoo, Diamantina, Hamilton, Georgina, Burke, Ranken, Warburton and Cooper Creek, carry the water through into Lake Eyre. Lake Eyre is special because it is still gradually sinking. It is one of the largest and most unspoiled desert river systems in the world, supporting a diverse range of wildlife. The basin itself covers one-sixth of Australia.

Analysing: The maps provide information about where and when significant bushfires and floods occurred in Australia between 2009 and 2016. They tell you how much land was impacted by fire and how many lives were lost during flood events.

Bushfires mostly occur in the southern areas of the country during summer when the weather is hot and dry and not in the more humid northern parts of Australia.

Communicating:

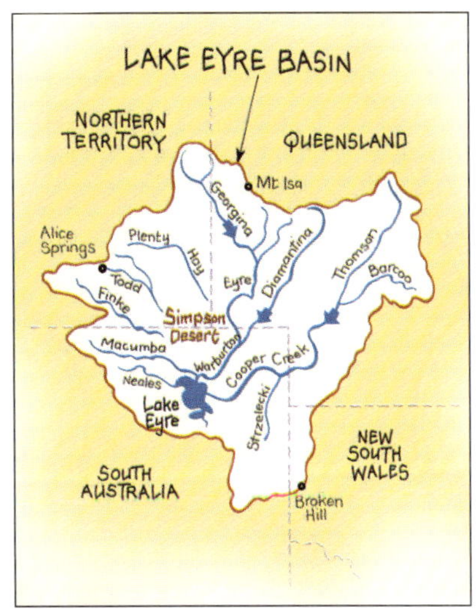

23 Reshaping Australia

Research:

2018 – Cyclone Marcus – category 5 – Northern Territory, Western Australia

2015 – Cyclone Marcia – category 5 – Queensland coast north of Rockhampton

2011 – Cyclone Yasi – category 5 – Northern Queensland near Mission Beach

2007 – Cyclone George – category 5 – Western Australia north-east of Port Headland

2006 – Cyclone Monica – category 5 - The Northern Territory (Arnhem Land)

2006 – Cyclone Larry – category 4/5 – Queensland near Innisfail

2005 – Cyclone Ingrid – category 5 – Cape York, Northern Territory Arnhem Coast, Western Australia Kimberly Coast

Questioning: Sample responses:

Where did you stay when your home was damaged? Were you injured? What did you lose? How long did it take to repair the damage? What has changed in your community since the event? How did you react to the event?

Analysing: There is no undergrowth or small shrubs to provide food and homes for animals. Bird nests and possum houses have been destroyed higher up in the canopy. The canopy is gone, so there is no shade. Water run-off will be polluted with ash and this will impact the water supply. There are no flowers for the insects to feed on and no insects means no food for birds or reptiles.

Communicating:

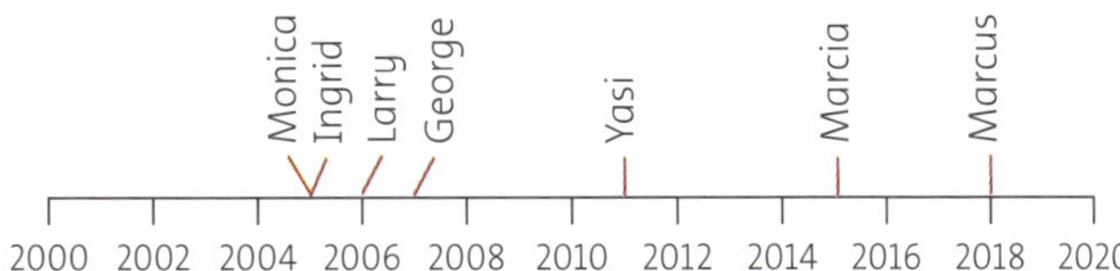

Evaluating and reflecting: Answers will vary, but include comments around the cost of repairing bushfire and flood damage, the impact on people's mental health, the fact that we might need to change where people live so they are not in flood or bushfire-prone areas, that we might need to change how buildings are constructed so they are fire resistant, communities might need to do more regular low-level burns to reduce fuel loads, communities might need to have less hard surfaces and more areas where the ground can soak up excess water.

24 Impacts of Bushfires

Questioning:

Banksias, Hakeas and Eucalypts use fire to help them reproduce.

The ash from fires adds nutrients to the soil.

Epicormic buds are small buds under the bark of a tree.

Disaster victims can recover from the trauma by speaking with counsellors and medical professionals.

TARGETING HASS 5 © PASCAL PRESS ISBN: 9781925726060

Answers

Note: Answers are not supplied to open-ended questions, where students' responses will vary.

Analysing:

Impact of bushfire	Actions taken after the fire
Burned houses and schools	Some people left and some rebuilt
Destroyed roads, bridges and powerlines	Infrastructure was repaired
Had to travel long distances to see a doctor or go shopping	Businesses reopening and new people moving to the town
Children had bad dreams	Speak to a counsellor at school

25 Australian Democracy and Values

Research: Sample answers:

The six Australian colonies were renamed as states. The states could now govern themselves in their own right as part of the Commonwealth of Australia. The Australian Constitution was written. A federal parliament was instituted. Each level of government (state and federal) was given power over a different area. The federal government was given control of taxation, defence, foreign affairs and communications (post and telephone).

Analysing: The trend in citizenship and voting is to become more inclusive over time, as women and, later, Indigenous peoples have been allowed to vote. Becoming an Australian citizen is now easier with the end of the *Immigration Restriction Act* (White Australia policy). The impact on society has been a positive one. It means that we have a fairer system of government for all Australians and there is less discrimination.

26 Let's Vote!

Research: The election advertising blackout rules start the Wednesday night before an election (polling day) and they stay in place until voting is complete at 6pm on Saturday night. This means that no political party or organisation can broadcast election advertising material on television or radio during the blackout period. This affects federal, state and territory elections. The blackout does not apply to online (internet) or print (newspaper and magazine) media. It also does not affect social media (Facebook, Twitter, Instagram, etc.).

Questioning:

statutory: permitted or enacted by statute (law decreed by the parliament)

referendum: a vote where everyone in the electorate votes on a particular nationwide proposal

electorate: the area represented by a Member of Parliament

ballot: a system of secret voting

marginal: referring to a seat in parliament that is held by a small majority and that is at risk in an election

Evaluating and reflecting: Sample answers:

Positive impacts of how-to-vote cards	Negative impacts of how-to-vote cards
Help people make decisions Minimise informal votes Minimise donkey votes	Can appear to be compulsory May take away freedom of choice from the voter Add to littering around polling booths

27 Enforcing the Law

Research: The role of the Australian Border Force is to protect Australia's borders and assist lawful trade and travel. They monitor people leaving and coming into the country at all airports and seaports, and check passports. They also check for illegal goods coming into the country such as food items, drugs and natural products that may contain insects harmful to Australia's ecosystems. During the COVID-19 pandemic, they helped to monitor international travel restrictions.

Analysing:

Sample response: It is difficult to regulate social interactions because individuals have their own thoughts and ideas and some people do not like to follow rules set by others. Some people believe that individual rights and freedoms should be maintained and not regulated by the government.

28 Getting Hands On

Questioning: Sample responses:

How do you know what to do for the different species of animals? Where do you keep animals when they are at your home? What do you feed them? When are animals released back into their native habitats? What happens if an animal cannot be released?

Analysing: Sample responses:

Adults are very busy with jobs and family so they don't have the time to get proper training. More people live in units/apartments, without the space to care for animals. The cost of looking after animals stops people from doing it. It is hard to volunteer time when the schedules for volunteer organisations are too inflexible. Not everyone is interested enough in the cause.

Evaluating and reflecting: Sample answers:

Making it easier and less time-consuming to do the training. Having certain times of the year when you can volunteer, and not having to be on call for whenever you are needed. Advertising the benefits of volunteering for yourself and the wider community. Advertising the different roles people can volunteer for, to increase awareness.

Increased volunteer rates would have a positive impact in society. Volunteering helps reduce stress and combats depression. It increases self-confidence and gives people a sense of purpose. Volunteering also helps develop a stronger sense of community and belonging.

Answers

Note: Answers are not supplied to open-ended questions, where students' responses will vary.

ANSWERS

29 How do I choose?

Questioning: Sample responses:

Why is that want important to you? How often do you want that product or service? Would you be willing to substitute it for a similar product or service?

Evaluating and reflecting: Answers will vary. If the response is that less choice is not better, possible reasons would be lack of individuality, lack of personal choice, boring, repetitive. If the response is that less choice is better, possible reasons would be making shopping quicker and easier, no arguing over which brand to buy, less confusion over product price and quality.

30 Resources and Food Production

Research: Answers will vary. Sample answers:

Food	Water used / unit
1 glass of milk	200 litres of water
1 egg	135 litres of water
1 apple	70 litres of water
1 potato	25 litres of water
1 kg rice	2497 litres of water
1 kg bread	1608 litres of water
1 kg beef	15 415 litres of water
1kg pork	5988 litres of water

Questioning: Answers will vary. Sample responses:

Why does producing beef use so much more water than producing pork? Which food product uses the most water to produce? Which food product uses the least amount of water to produce? How can the water used to produce food be reduced?

Analysing: Chocolate production can also be wasteful because the outer husk of the pod is thrown away instead of being used in another product. The shells from the seeds are also 'blown away' and not used or reused. Chocolate also comes in a lot of packaging, some of which may not be needed.

Communicating:

Chocolate production flow chart

Cacao plants grow 3–4 years

⇩

Cacao fruit harvested by hand

⇩

Fruit pods split open, beans removed

⇩

Bean covered in banaa leaves and fermented for 8 days

⇩

Beans dried for 1–2 weeks

⇩

Dried beans taken to a factory

⇩

Beans sorted, cleaned and roasted

⇩

Nibs ground into a paste and squeezed to form cocoa butter

⇩

Cocoa butter mixed with milk and sugar to make chocolate.

31 Use of Resources

Research: Natural resources are materials or substances that occur in nature, without the actions of people. Examples of natural resources are: water, coal, oil, natural gas, iron, minerals, timber. Capital resources are goods produced and used to make other goods and services. Examples of capital resources are: tools, equipment, machinery, buildings.

Questioning: Sample responses:

regrowth: What happens to plant life after a bushfire?

natural cycles: What are the patterns of seasons and the lifecycles of animals known as?

interdependence: Which word describes the way in which plants and animals work together to benefit each other?

seasonal: What clues helped Aboriginal peoples decide to move from one area to another so that food supplies could be maintained at a sustainable level from year to year?

Analysing: Natural resources: animals, plants, fruits, vegetables, the environment. Aboriginal peoples used natural resources sustainably because they never used up the supply of any plant or animal in an area. They took only what they needed and they didn't waste anything.

Capital resources: firesticks, tools used for hunting animals and preparing medicines. These capital resources were made from natural resources and were only made when required. They were not thrown away, but reused or repurposed if damaged.

Evaluating and reflecting: Answers will vary. Sample responses may include:

Yes we can learn from Indigenous land management methods. The most important lesson would be to avoid unnecessary waste and to not overuse resources in a particular place or environment. We need to change our modern behaviours by being more careful about our use of natural resources and making things last instead of using items with a short lifespan.

32 Following Fads

Research: Answers will vary. Sample responses: beyblades, taking selfies, angry birds, pokemon, cupcakes, dabbing, Dubstep, fidget spinners, homemade slime, rainbow loom.

Questioning:

1. False, 2. True, 3. False, 4. True, 5. False, 6. True

TARGETING HASS 5 © PASCAL PRESS ISBN: 9781925726060

Answers

Note: Answers are not supplied to open-ended questions, where students' responses will vary.

History Assessment 1

Discovery of gold in Coolgardie

It is a photograph.

It is primary source material.

Its purpose is to show us life on the goldfields.

The image is black and white. The image looks busy, there is lots of equipment, most of it made from iron and timber laying around. The men have broad floppy hats on and are wearing simple trousers and shirts. They have large dishes for washing the crushed rock in to look for gold. There are a few trees in the background, but the area of the diggings has no plant life.

The image tells you about the tools and equipment used when prospecting for gold, what clothing people wore, how they interacted with the environment and how they worked – most of the work was done by hand and there weren't many machines to assist in the gold-mining process.

The image is trying to convey a feeling of how much hard work and physical labour went into mining for gold. It also suggests that a lot of effort was put in for little return as most of the dishes are empty.

To better understand the image, you would need to know how big their mining plot of land was, how long they had been mining for, and how much gold they had found so far.

Back Country Squatter

It is cartoon.

It is primary source material.

The purpose is to show the situation facing squatters in the 1800s and all the problems they face.

The image is brown and white. It shows the squatter wading through a sea of debt, showing that he owes lots of money. He is carrying a heavy load of lots of problems, with the introduction of rabbits topping it off. He is on the road to ruin, as shown by the road signpost. There is not a lot of plant life growing in the area. He is not well dressed and his back is bent, which means he is feeling the pressure of the situation.

The image tells you that life in colonial times was harsh. It wasn't easy to make money and people had to face lots of difficult situations.

The feeling is one of sympathy. We feel sorry for the squatter and we are trying to understand the problems facing him.

To better understand the image, you would need to know what part of the colony he was living/working in and if he had a wife and family. (Additional answers here may vary.)

History Assessment 2

Overlander explorers: Some of the first overlanders were Joseph Hawdon, John Gardiner and John Hepburn. They drove cattle and sheep long distances from station to station, or from the station in the country to the markets and ports near the capital cities. For this reason, they were commonly known as drovers. Overlanders had to be strong and independent, excellent horsemen and able to manage the cattle or sheep as well as the other drovers. The most famous overland stock routes were the Birdsville Track, the Strzelecki Track and the Canning Stock Route that runs from Halls Creek in the Kimberly region of Western Australia to Wiluna in the mid-west region of the state. With the development of the railways, the need for overlanders declined. They were replaced in the mid 1900s with road trains.

Bushrangers: Legends show Australia's bushrangers as dashing folk heroes who were unfairly persecuted by the police. Some may have been, but others were simply escaped convicts or petty thieves who turned to robbing innocent settlers and stealing the proceeds of their hard labour. Many of the bushrangers in the 1860s and 1870s were the sons of ex-convicts and had been born in Australia. They had excellent bush skills and were good horse riders as they had grown up on farms and around livestock. Bushrangers typically disliked authority and did not respect the law. Many of the poor settlers who were struggling identified with the bushrangers and supported them. Some of Australia's most notable bushrangers were 'Mad' Dan Morgan, 'Brave' Ben Hall, and Ned Kelly and the Kelly Gang, who had a shootout with police at the Glenrowan Hotel in 1880.

Native police: The first attempt at establishing a Native Police Corps occurred in 1837 in Port Phillip (originally part of colonial New South Wales but now part of Victoria). The main goals were for Aboriginal troopers to be 'civilised' role models for other Aboriginal peoples, and to capture runaway convicts. However, the attempt was a failure and the Corps was disbanded. Queensland's Governor Charles Fitzroy established its native police in 1848. It was led by Commandant Frederick Walker. It was made up of several small detachments of Aboriginal troopers led by European officers. They violently put down any Aboriginal resistance to colonial rule. In 1880, Queensland native police trackers were sent to Victoria to assist in the hunt for the bushranger Ned Kelly. The Queensland Native Police Force was disbanded in 1897.

A

Answers

Note: Answers are not supplied to open-ended questions, where students' responses will vary.

ANSWERS

Chinese goldminers: When gold was discovered in the colony in the 1850s, Chinese immigrants came, hoping to strike it rich. Over 40 000 Chinese people (mostly men) arrived in Australia between 1852 and 1889. Chinese gold diggers started gardening and growing fruits and vegetables on the goldfields. They set up 'cookshops' that attracted European diggers to their camps. After the gold rush, Chinese settlers worked in market gardens or on farms. Many set up small grocery stores and became merchants. There was a lot of tension between Chinese migrants and European miners on the goldfields. This led to many conflicts, including the Lamming Flat riots. From the 1860s, government legislation, the *Chinese Migration Act*, was used to restrict Chinese migration. The premier of NSW, Henry Parkes, also said that British and Chinese people should not marry each other.

History Assessment 3

Sample response:

1788: European settlement.

1790: The second fleet arrives.

1791: The third fleet arrives.

1803: Matthew Flinders circumnavigates Australia, proving that it is an island.

1808: the Rum Rebellion and Governor Bligh is removed from office.

1813: The Blue Mountains are crossed by Blaxland, Lawson and Wentworth, opening up the western regions of NSW.

1824: The penal colony at Moreton Bay (now Brisbane) is established.

1829: Perth is founded on the south-west coast of Western Australia.

1833: The penal settlement of Port Arthur was established in Van Diemen's Land (now Tasmania).

1835: The settlement of Port Phillip (later becoming Melbourne) is created.

1836: South Australia proclaimed.

1843: The first elections are held for parliament.

1851: Gold is discovered in Victoria.

1854: The Eureka rebellion.

1859: Queensland separated from NSW.

1868: Britain stops sending convicts to Australia.

1872: The overland telegraph opened, linking Darwin and Adelaide.

1880: The bushranger Ned Kelly is hanged.

1883: The Sydney to Melbourne railroad is opened.

1890: The Australian federation conference called a constitutional convention.

1900: The British parliament passed the *Commonwealth of Australia Constitution Act.*

Geography Assessment 1

1. Before the flood, people would have used the waterway for transportation and recreational activities. They would have driven along the roads, walked along the footpaths that had been created by changing the natural environment. The people living along the edge of the river would have been influenced by its tides, which is why their houses are set back a few metres from the edge of the river. Floodwater has covered all land-based access areas such as roads and paths. People would be unable to use the spaces as they normally would. The water would be polluted and so even normal recreational water activities would stop during and immediately after the flood.
2. The changes made by people include the building of roads, footpaths and driveways. These hard surfaces are non-porous and do not let the water soak into the ground. This would have made the floods more severe. The construction of houses along the riverbank also increased the severity of the flood's impact. If houses were not so close, fewer of them would have been damaged by the rising water levels. People have also cut down native trees from the area along the river (for home and infrastructure construction). Less plant cover means that during times of flood, there are greater levels of soil erosion.
3. As a result of the floods, people could rebuild their houses and raise them higher (using concrete slabs or stilts). They could improve the drainage along the roadways so that high volumes of water can escape more easily. They could plant more trees and shrubs to limit erosion.

Geography Assessment 2

1. Its distance from other inhabited areas; its geographical isolation from other continents; its extreme low temperatures; its lack of infrastructure, which means that rubbish and waste cannot be processed there; in winter and autumn it is always dark.
2. Historical whaling and seal hunting; pollution leading to climate change and increasing global temperatures; exploration which led to equipment and supplies being simply left in Antarctica; people working in Antarctica have produced sewerage and spilled oil and petrol; people visiting Antarctica have brought in non-native plant and animal material to the area on their shoes and clothes.
3. Taking waste to other countries for disposal; limiting the number of tourists and scientists visiting each year; signing the Antarctic agreement to agree on how to manage and protect the area.
4. Increased melting of the ice; destruction of the habitats of Antarctic animals such as penguins, leading to their possible extinction; damaging the food chain of animals that migrate to the Antarctic waters, e.g. whales; arguments between countries over the area and its environment.

TARGETING HASS 5 © PASCAL PRESS ISBN: 9781925726060

Answers

Note: Answers are not supplied to open-ended questions, where students' responses will vary.

Geography Assessment 3

1. NSW, Vic., Tas.
2. High temperatures, low humidity, low rainfall, high winds
3. Fire helps to open seed pods and acts as a trigger for germination. The ash can increase soil fertility. Fires also clear away thick understorey plants, encourage new growth which then provides food for animals.
4. Some trees and forest areas can take hundreds of years to recover. Fires kill native animals and livestock on farms. Fires destroy the homes of animals such as nesting birds and creatures that live in hollow logs. Bushfires can lead to increased soil erosion and increased sedimentation of creeks and rivers.
5. Sample response:

Step 1 – Action: Discuss what to do with your family. **Reason:** So everyone knows what to do and there is less confusion at the time of the emergency.

Step 2 – Action: Prepare your home by mowing the grass, trimming overhanging tree branches and not having bushes too close to the house and fences. **Reason:** So there are less opportunities for fire to get a hold near your home.

Step 3 – Action: Prepare your home by clearing leaves from gutters on the roof and having hoses and a supply of water nearby. **Reason:** If embers get into the gutters, they can quickly spread fire to the roof.

Step 4 – Action: Move gas cylinders away from the side of the house. **Reason:** These can explode if they get too hot, even without the direct impact of fire.

Step 5 – Action: Use your state's bushfire app to keep up to date with nearby fires and alert levels. **Reason:** Local conditions can change quickly. It is important to stay up to date so you are aware of what is happening and you can respond quickly.

Civics and Citizenship Assessment

1. Answers will vary.
2. These rules do not match with Australia's democratic values. They are not fair or equal for all. They discriminate against certain people. They do not allow people freedom.
3. Sample response: I would organise a petition of the students and teachers to try and get the rules changed. I would organise a peaceful protest with banners to complain about the new rules. I would write a letter to the principal to try and convince them to change the rules.
4. Sample response: This would be a good idea because my actions would be peaceful and they would try and get as much community support as possible.
5. Answers will vary.
6. Answers will vary.

Economics and Business Assessment

1. It is mostly a service, because no physical goods are transferred between the seller and the buyer.
2. It is a want, because it is not needed for survival. It is something that is desired.
3.

Human resources	Natural resources	Capital resources
Person to do the massage Musician to play/create music	water flowers	Banking services for payment Devices such as a computer and speakers for playing music

4. The strategies include calming colour tones, appealing directly to the reader, using personal pronouns, highlighting the negatives in consumers' lives at the moment, offering group discounts, 'internationally renowned' claim, using rhetorical questions.
5. Answers regarding effectiveness of the advertisement will vary.
6. Sample responses: Do they have enough money to spend on a visit to the day spa? What will they miss out on (trade-off) in order to use their money at the day spa? How scarce are visits to a spa like this? Is it something that the consumer will really want to do because it is rare and therefore special?

Targeting HASS
Year 5

Copyright © 2020 Pascal Press
ISBN: 978 1 925726 06 0
Reprinted 2022, 2025

Published by Pascal Press
PO Box 250
Glebe NSW 2037
contact@pascalpress.com.au

Author: Merryn Whitfield
Publisher: Lynn Dickinson
Typesetter: Ruth Schultz
Editor: Ruth Schultz
Designer: Janice Bowles

Printed by Vivar Printing/Green Giant Press